MORE PRAISE FOR *THE KNOW-IT-ALLS*

"These finely researched portraits are a joy."   *Nature*

"Individualism is a big part of what makes America great—until it becomes a euphemism for selfishness and arrogance among lucky winners who prefer to believe that luck and other people had nothing to do with their success. *The Know-It-Alls* is a terrific case study of some of the unreckoned costs of the digital revolution, and how one piece of the American idea threatens to overwhelm the others."

Kurt Andersen, author of
Fantasyland: How America Went Haywire

"Noam Cohen's *The Know-It-Alls* provides a provocative and illuminating examination of Silicon Valley. Using profiles of its core digital capitalist giants and the immense political, economic and cultural power they have quickly come to possess, Cohen raises troubling questions about how this can possibly square with a fair, decent, humane, and democratic society. This immensely readable book should be mandatory reading."

Robert W. McChesney, author of *Digital Disconnect*

"A fascinating intellectual profile of the people who have increasingly come to rule our world. With precision and skill, Noam Cohen tweaks the pretensions of a handful of tech oligarchs, whose self-styled project to better our lives results in little more than a power grab at our economy and our democracy. As America's center of gravity inexorably shifts to Silicon Valley, and the original vision of a decentralized Internet of personal expression gets drowned in a sea of commerce and advertising, I'll be turning to Cohen's insights into the profiteers responsible again and again."

David Daven, author of *Chain of Title*

"An enli   naires have
shifted p   *Kirkus*

THE KNOW-IT-ALLS

The Rise of Silicon Valley as a Political
Powerhouse and Social Wrecking Ball

NOAM COHEN

ONEWORLD

A Oneworld Book

First published in Great Britain and Australia by Oneworld Publications, 2018
This paperback edition published 2019

Published by arrangement with The New Press

ISBN 978-1-78607-490-4
eISBN 978-1-78607-368-6

Typeset by Bookbright Media
Printed and bound in Great Britain by Clays Ltd, Elcograf S.p.A.

Oneworld Publications
10 Bloomsbury Street
London WC1B 3SR
England

Stay up to date with the latest books,
special offers, and exclusive content from
Oneworld with our newsletter

Sign up on our website
oneworld-publications.com

MIX
Paper from
responsible sources
FSC® C018072

Why, anybody can have a brain. That's a very mediocre commodity. Every pusillanimous creature that crawls on the Earth or slinks through slimy seas has a brain.

—*The Wizard of Oz*

The end of man is knowledge, but there is one thing he can't know. He can't know whether knowledge will save him or kill him. He will be killed, all right, but he can't know whether he is killed because of the knowledge which he has got or because of the knowledge which he hasn't got and which if he had it, would save him.

—*All the King's Men*

CONTENTS

THE KNOW-IT-ALLS

INTRODUCTION

"To Serve Man"

I n a memorable *Twilight Zone* episode, "To Serve Man," aliens land on Earth. These aliens, the Kanamits, nine feet tall and topped with massive heads, say they've come in peace and intend to share their superior technology to benefit humanity. Immediately, they are true to their word. Barren soil in Argentina produces grain; mysterious force fields protect each nation's borders, rendering the nuclear arms race irrelevant. And when the suspicious Soviets raise concerns, the Kanamits' chief gladly takes and passes a lie detector test. A little while later, when the aliens suggest that Earthlings load up in a flying saucer to see the wonders on the Kanamits' home planet, few question it. There are lines to get a precious seat.

The story is told in flashback through the eyes of one of those passengers, Michael Chambers, an American code breaker assigned to decipher a manuscript accidentally left behind by the Kanamit leader. A member of Chambers's decryption team succeeds in piecing together the manuscript's reassuring title, "To Serve Man," and the world is confirmed in its belief that the aliens' intentions are good. Chambers rushes onto the Kanamit bandwagon as one of the last passengers aboard. Yet just as Chambers walks up the ramp of the aliens' ship, a voice below reveals the bitter truth about "To Serve Man": "It's a cookbook!" At the end, a Kanamit is heard over a loudspeaker encouraging Chambers to be sure to eat all of his supper.[1]

This story is flamboyantly absurd science fiction: How can you crack a code without a key to work off of? And would the Kanamit language really have the exact same double meaning for the phrase *to serve*? Furthermore, why would aliens come all this way to harvest people instead of something truly tasty like cattle or tuna or truffles? "To Serve Man" nonetheless manages to convey an important message: it is wise to be suspicious of those who claim to pursue selflessly the prosperity of others even as they pursue their own. Also, those dual meanings of *serve* may reveal a universal truth, in that purporting to act in service of others without their consent necessarily involves manipulation, grooming, and exploitation.

Silicon Valley surely is unrivaled in the American economy in its claims to "serve mankind." So much so, in fact, that the satirical TV show *Silicon Valley* has a running joke that whenever a start-up founder is introduced, no matter how absurdly technical his project may be, he assures the audience that he is committed to "making the world a better place." Paxos algorithms for consensus protocols . . . making the world a better place.[2] Minimal message-oriented transport layers . . . making the world a better place.[3] Yet strip away the satire, and you find that Google works from the same playbook. The company assures us that it collects and stores so much personal information about its users to better serve them. That way, Google sites can remember what language you speak, identify which of your friends are online, suggest new videos to watch, and be sure to display only the advertisements "you'll find most useful."[4] Even when Google is being paid by businesses to show you ads, it's really thinking about making your life better!

Facebook similarly insists that it acts in the best interest of humanity, no matter how its actions may be perceived. For example, there is the Free Basics project, which provides a Facebook-centric version of the Internet for cell phone users who cannot afford access to the actual Internet.[5] Critics in India objected to Facebook's apparent largesse, seeing the program as pushing a ghettoized, fake-Internet experience for poor people merely to keep its audience growing. Facebook's chief executive, Mark Zuckerberg, didn't back down, however, describing the dispute as a choice between right and wrong, between raising hundreds of millions of people out of poverty through even limited Internet access or leaving them to suffer without any access at all. He made a public appeal by video, which concluded, "History tells us that helping people is always a better path then shutting them out. We have a historic opportunity to improve the lives of billions of people. Let's take that opportunity. Let's connect them."[6]

Certainly, from time immemorial, moguls have believed that their own prosperity must be good for all of society, but only the recent batch of Silicon Valley entrepreneurs have acted as if money were an unanticipated byproduct of a life devoted to bettering mankind. Marc Andreessen, the Silicon Valley venture capitalist who serves on Facebook's board, was scathing when he learned that the Indian government had sided with the critics and blocked Free Basics. The government's decision was "morally wrong" and punishing to the world's poorest people, Andreessen wrote on Twitter, offering yet another example of how India has been on the wrong track since its people kicked out their British overlords. "Anti-colonialism has been economically catastrophic for the Indian people for decades. Why stop now?" he asked sarcastically. Andreessen quickly apologized when he saw the furious response to those comments, particularly within India,[7] but they nonetheless proved that he belonged among a tiny class of public figures who would have the self-assurance to make such a statement in the first place, to trash Indian democracy and self-determination in defense of their own belief systems and their own particular business models.

The Know-It-Alls is the story of these powerful, uber-confident men, starting with Andreessen, who helped nurture the World Wide Web to prosperity in the 1990s before switching to investing. It ends with Zuckerberg, who has the most ambitious plans for linking the world within his own commercial online platform. Along with Andreessen and Zuckerberg there's a bevy of Internet billionaires, including Jeff Bezos of Amazon, Sergey Brin and Larry Page of Google, Reid Hoffman of LinkedIn, and the early Facebook investor Peter Thiel. They are a motley crew—some, like Hoffman, are outwardly friendly, cuddly even, while others, like his good friend Thiel, cultivate an aura of detachment and menace. Some, like Brin and Page, one suspects would prefer to be left alone with their computers, while others, like

Bezos or Zuckerberg, seek the limelight. Some were born to program, others to make money. But they share common traits: each is convinced of his own brilliance and benevolence, as demonstrated by his wildly successful companies and investments, and lately each is looking beyond his own business plans to promote a libertarian blueprint for us all.

Collectively, these Silicon Valley leaders propose a society in which personal freedoms are near absolute and government regulations wither away, where bold entrepreneurs amass billions of dollars from their innovations and the rest of us struggle in a hypercompetitive market without unions, government regulations, or social-welfare programs to protect us. They tap into our yearning for a better life that technology can bring, a utopia made real, yet one cannot escape the suspicion that these entrepreneurs may not fully appreciate what it means to be human. That is, not just to be a human individual—the unit that libertarianism is so obsessed with—but to be part of a family, a community, a society.

The feminist political theorist Susan Moller Okin argued convincingly that libertarianism requires precisely this kind of obtuseness. In the libertarian fantasy, men magically arrive at adulthood ready to remake the world: How? Raised by whom? If advocates for extreme individualism actually had to acknowledge the work and sacrifice of women to bear and nurture children, Okin contended, as well as the assistance of society in children's upbringing, their arguments would lose all force. No one would then be able to say with a straight face that whatever he has is the product of his own hard work and should be his alone to control. "Behind the individualist façade of libertarianism," she concluded, "the family is assumed but ignored."[8]

Once women, family, and society are pushed to the side, however, individuals are free to duke it out for life's spoils unencumbered by social obligations, as Hoffman explains in his business advice book *The Start-up of You*. "For anything desirable, there's competition," he writes. "A ticket to a championship game, the arm of an attractive man or woman, admission to a good college, and every solid professional opportunity." The only sensible response, he concludes, is to labor as a high-risk, high-reward "start-up of you"[9]:

The conditions in which entrepreneurs start and grow companies are the conditions we all now live in when fashioning a career. You never know what's going to happen next. Information is limited. Resources are tight. Competition is fierce. The world is changing. And the amount of time you spend at any one job is shrinking. This means you need to be adapting all the time. And if you fail to adapt, no one—not your employer, not the government—is going to catch you when you fall.[10]

As the harsh world dreamed up by these wealthy, powerful Silicon Valley leaders gains traction, *The Know-It-Alls* becomes the story not just of their lives but of ours, too.

Silicon Valley never would have had the wealth and power to shape America's values had there not been a World Wide Web to make computers so useful and relevant to daily life. When the British physicist Tim Berners-Lee first brought the Web into existence some twenty-five years ago at the CERN laboratory in Switzerland, he imagined he was creating a decentralized network for people to collaborate through their computers, with commerce low among his priorities.[11] Berners-Lee's original vision of a small-scale, almost anarchic Web was shed nearly immediately as Netscape, the Silicon Valley company Andreessen cofounded after graduating from college in the Midwest, took the lead in the Web's development. Netscape's early emphasis on commerce and creating a passive, user-friendly experience led the Web to where it is today—wildly popular around the world, with a few companies able to apply a chokehold on how we access and use the Internet. In search, there is Google. In commerce, Amazon. In social networking, Facebook.

Yes, despite its European parentage, the Web would be a Bay Area baby. Vital new tools for navigating within and between sites, for searching through oodles of digital information, and for sharing opinions and photos with friends and acquaintances all grew to maturity within Silicon Valley's start-up culture. Businesses based on those tools soon directed a sizable portion of the nation's wealth toward the West Coast, as if the United States were a pool table tilted so the balls wound up in the left side-pocket. The wealth that has since accumulated around San Francisco has largely gilded over its "flower power" reputation, leaving the city inhospitable to anyone but the highest-paid programmers, who are shuttled to and from their corporate campuses on luxury buses.[12] Street protests against those buses, which serve as a private mass transit system, have helped highlight the great wealth disparity in the Bay Area, but there have been other extravagances as well. The tech investor Sean Parker staged a multimillion-dollar wedding in a redwood forest landscaped to look like Middle-Earth of *The Lord of the Rings*.[13] The Silicon Valley venture capitalist Vinod Khosla caused an outcry by demanding that the state pay him $30 million before he would give the public access to Martins Beach, which sits below his property.[14] And there was the lavish, Versailles-themed fortieth-birthday party in Los Angeles for David Sacks, a former PayPal executive and successful start-up founder, who to his credit at least tried to stop his guests from sharing the gaudy details through social media.[15]

The consequences of Silicon Valley values went from classless to catastrophic, however, during the recent presidential election. A near-majority of the electorate succumbed to Donald Trump's appeal to bring back a less convulsive past, complete with its unchecked racism and misogyny, and many of us experienced for the first time the fragility of our society after so much Internet-based "disruption." America in 2016 lacked the stabilizing influences of traditional news-gathering organizations and community groups, vibrant local businesses, strong labor unions, aggressive government regulations, and engaged political parties, each of which had been undercut by Silicon Valley businesses and the libertarian principles of their founders. What remained were a few distant tech giants and a collection of angry individuals, abandoned by the global economy and lashing out at remote forces—immigrants, Wall Street bankers, trade agreements, political correctness—without serious intent. Instead, these voters empowered a cynical blowhard who promised, improbably, "I alone can fix it."

Among the circumstances for Trump supporters to rebel against were the Silicon Valley insta-billionaires themselves, who had helped bring about so much of the country's social disruption. The rapid rise of these young entrepreneurs sent an unmistakable signal that income inequality would only be getting worse. At the same time, the apparent requirement that a successful entrepreneur attend the right school and have the right backers revealed that the Silicon Valley start-up system wasn't a meritocracy, as is so often proclaimed, but was rigged, to quote the great man himself. Take the case of the photo-sharing service Instagram, which was sold for $1 billion to Facebook barely two years after being launched. One cofounder, Kevin Systrom, a twenty-eight-year-old Stanford graduate, kept 40 percent of the proceeds of the sale, with a few prominent VC firms and early investors taking big cuts as well.[16] At the time of the sale, in April 2012, Instagram employed all of thirteen workers. How any of this could help sustain a happy, productive society was a mystery.

There was a final gift from Silicon Valley during the 2016 election: the radical insistence that what was expressed on the Web should be unregulated, which allowed the hate and abuse of the Trump campaign to fester and then spread. On Twitter, Trump's followers and Trump himself were permitted to intimidate critics, particularly women and minorities.[17] This relaxed approach from Twitter was matched by Facebook and Google, which served their users made-up news about the election as long as the articles remained popular. Freedom of speech apparently trumped all other values as Google, Facebook, and Twitter encouraged the public to stew in their own hateful juices and profited handsomely in the process.

One Silicon Valley figure unafraid to explore the natural affinity between Silicon Valley values and Trump values is Thiel, who saw Trump as a Silicon Valley–style man of action and vision, a larger-than-life agent of disruption. "When Donald Trump asks us to Make America Great Again, he's not suggesting a return to the past," Thiel explained in his speech to the delegates gathered at the Republican National Convention in Cleveland. "He's running to lead us back to that bright future." That was the point, wasn't it? To apply the winning, destructive, forward-thinking vision of Silicon Valley to the rest of America. As Thiel boasted in that same address, "Where I work in Silicon Valley, it's hard to see where America has gone wrong. My industry has made a lot of progress in computers and in software, and, of course, it's made a lot of money. But Silicon Valley is a small place. Drive out to Sacramento, or even just across the bridge to Oakland,

and you won't see the same prosperity. That's just how small it is."[18] Imagine if everything in the American economy worked like Silicon Valley! This was the glorious future Thiel saw in a Trump presidency.

Thiel's high-profile endorsement of Trump certainly raised the hackles of his peers, who generally supported Hillary Clinton for president, seeing her as the continuation of the Obama administration's Silicon Valley–friendly policies on immigration and Internet regulation. But Thiel was also an outlier for being so high-profile in his support during the election, which included $1.25 million in donations to a Trump-affiliated super PAC and the Trump campaign itself.[19] The founders of Internet start-ups, like entertainers or professional athletes, aspire to be popular with all sorts of people and are quick to play down political differences. They claim to be focused on efficiency, not ideology. Elon Musk, who started as a Web entrepreneur before founding the electric car company Tesla, captured this nonpolitical political perspective in a post to Twitter in 2012: "I'm neither anti-conservative nor anti-liberal. Just don't like group think. Ideas should be considered on their own merits."[20] Even Thiel himself later sought cover from some of Trump's more extreme ideas—a wall with Mexico, mass deportations—by saying he and other supporters took Trump "seriously, but not literally."[21]

The libertarian tilt of the Know-It-Alls has been of great assistance as they pursue a version of nonpolitical politics. Libertarianism can be framed as moderate and open-minded: that is, I agree with liberals on some issues like gay rights or abortion rights, but agree with conservatives on others, like tax cuts or shrinking the social safety net. Similarly, the libertarian can say even-handedly that though the left may hate it, he believes in absolute freedom of speech, and though the right might hate it, he believes in letting people smoke marijuana if they want to. This approach fits someone like Jeff Bezos, for example, who has donated to a campaign to legalize gay marriage in Washington State as well as one to defeat a ballot initiative that would have introduced an income tax on millionaires in the state. Bezos, who now owns the *Washington Post*, has also supported the foundation that publishes the libertarian magazine *Reason*. Some observers have labeled Bezos a "liberaltarian," a liberal libertarian, which is a term that could apply to many Silicon Valley leaders, who travel in Democratic Party circles but oppose unions, hate-speech codes, or expanded income redistribution.[22]

And isn't this the rub, really, of any book trying to explain the political influence of Silicon Valley leaders? So much of what they are advocating comes at you sideways or is described not as a belief but as an inevitable turn as society matures technologically. Yet there is, of course, a distinct Silicon Valley belief system. As we've seen, it advocates for a highly individualistic society led by the smartest people, who deliver wonderful gadgets and platforms for obtaining goods, services, and information efficiently, freeing each of us to compete in the marketplace for our daily bread. There is a particular history, too, of how those values came to be, which reflects separate but intertwined influences. First, there were the original hackers of university-run computer labs, a boys' club of programming geniuses who were a source of the optimism and idealism of Silicon Valley as well as its suspicion of authority and unwelcoming attitude toward women. Later came the entrepreneurs and investors congregating around Stanford University, who were early to recognize the windfall from computers once they had been improved so that ordinary people could use them. Silicon Valley's investors and entrepreneurs taught the hackers to think of the people who used their products as assets to extract value from, rather than simple folk who through the kindness of programmers would learn about the infinite power of computers.

The hackers arrived on the scene at the Massachusetts Institute of Technology in the late 1950s, when they were first introduced to computers by a pair of junior mathematics professors, John McCarthy and Marvin Minsky. Barely in their thirties, those two had already helped chart a path toward artificial intelligence through computers, which they believed could be programmed to think as people do. McCarthy taught MIT's first freshman class in programming in 1959, and those students naturally gravitated to the computer in McCarthy and Minsky's well-funded lab, where they were granted an extraordinary level of independence, freedom, even omnipotence.

These students came to be known as "hackers" because of how much they had to figure out on their own: when problems cropped up, they had to "hack together" a solution. Surrounded by their beloved machines, which seemingly only they truly understood, the hackers were permitted by McCarthy and Minsky to live by a set of anti-authoritarian rules that made sense for a bunch of smarty-pants outsiders. The individual outranked the collective. Personal freedom was more important than empathy or compassion. Status came from programming skill, not age or grades or likability or some academic title. In sum, the hackers believed in an ethic that gave each individual the freedom to do what he wanted with his computer and to say whatever he wanted about whomever he wanted whenever he wanted. Success or failure would be based on talent alone. Still, for all the acceptance of personal eccentricity and insistence on merit, these young men reflected a uniformity that persists in Silicon Valley to this day, starting with the fact that they were all men: women weren't exactly forbidden to be hackers, they just weren't accommodated or made to feel welcome, and at times they faced harassment.[23]

The other source for Silicon Valley's values, the tech entrepreneurs, were spurred on by Stanford, which by the 1950s had turned itself into an explicitly pro-business research institution. The university was founded back in the late nineteenth century with a robber baron's multimillion-dollar bankroll, yet for much of that history it lagged behind the great institutions back east. By the mid-twentieth century, Stanford was stumbling along, known for "educating the children of the middling rich of Los Angeles."[24] During this mediocre era, the school's ambitious engineering school dean, Frederick Terman, was given broad powers as university provost and vice president to make Stanford great. A specialist in a highly practical aspect of electrical engineering, radio waves, Terman had a proven record of turning research into business opportunities, and his plan was built on that experience: he proposed that Stanford use its resources to encourage research in areas with practical applications so that students and faculty members could help industry thrive. Surely a significant share of the wealth and status they generated would find its way back to Stanford.

Under Terman's guidance, Stanford smoothed the way for researchers to partner with big business or to strike out on their own. "I used to go around and give talks to people in industry," he recalled. "My theme was always that the university is a real asset if you make use of it—industrial use. And then I would come back and beat on the backs of the professors to get out and get acquainted with those companies that were related to their research."[25] An early example of a university-brokered business success was Hewlett-Packard, which was founded in 1939 by two of his favorite electrical engineering students, William Hewlett and David Packard. Terman, who was a professor at the time, helped Hewlett and Packard obtain their first equipment, which was initially housed in a Palo Alto garage that today is listed in the national registry as the "birthplace of Silicon Valley." He then helped them land their first large order, which was to produce equipment used for the sound editing on the Walt Disney movie *Fantasia*.[26] Hewlett and Packard in turn became prominent donors to Stanford. The culmination of their generosity toward Stanford came in 1977, when they raised the funds for the $9.2 million Frederick E. Terman Engineering Building.[27]

Terman ultimately was instrumental in Stanford's decision to invest in the new field of computer science, but at first he was skeptical. The business potential wasn't obvious. In the mid-1950s, computers were just fast calculators that were helpful to applied mathematicians but to few others. The research into artificial intelligence by McCarthy, Minsky, and others, however, helped make the case for computers' broader relevance. These professors—and particularly the obsessed young hackers who worked, ate, and slept in their labs—were pushing computers to do more, creating new programming languages, devising smarter hardware designs, and proposing outlandish challenges, like playing chess against humans. Could robot butlers or robot soldiers or automated translators be far behind? Just think of the business potential then.

In 1962, Stanford successfully recruited McCarthy with an offer of full tenure and better weather. That same year, he started work on re-creating his artificial intelligence lab as SAIL (Stanford Artificial Intelligence Laboratory), and in January 1965 he was one of four founding members of the Stanford computer science department, among the first in the country.[28] McCarthy spent much of his time at SAIL increasing students' access to computers, confident that the more computers and humans interacted, brain to brain, the more each could learn from the other. His lab was a pioneer in using monitors and keyboards to allow many individuals to communicate simultaneously with a central computer, a breakthrough he called "time-sharing."

Though McCarthy never cashed in on this work—research and commerce were in conflict, he believed—his lab was the fountainhead for a wide range of Silicon Valley companies, including Sun Microsystems, Silicon Graphics, Cisco, and Yamaha's music synthesizer business.

For all that success, however, the Stanford model of integrating the hacker and the entrepreneur only fully flourished with the widespread adoption of the World Wide Web starting in the late 1990s. Richard Weyhrauch, who spent the 1970s as a researcher at the Stanford Artificial Intelligence Lab, recalled that students and faculty members back in his day would sometimes leave to start companies, but they knew that the stakes were lower. "When we were in the A.I. lab, nobody would have thought you could build a company with more than a billion customers," he said.[29] By the time the Web made such a dream of global domination plausible, a thriving venture capital industry had grown nearby, along Sand Hill Road in Menlo Park, and Stanford was reliably feeding tech entrepreneurs into the system. Thus, before the young graduate students who created the Google search engine, Sergey Brin and Larry Page, had even set up a for-profit business they were handed a $100,000 check by an interested investor. This anticipatory investment was brokered by a university computer science professor who worked down the hall from Brin and Page and was himself already a wealthy tech entrepreneur. In less than a year, Google had incorporated and concluded a $25 million investment round with two Sand Hill firms.[30]

In the subsequent two decades, Stanford has flourished, becoming arguably the most desired college in the country, granting admission to fewer than 5 percent of applicants and earning the nickname "Get Rich U." from the *New Yorker* because of the success of its students.[31] Indeed, anyone pointing out the serious flaws with the world Silicon Valley is constructing must accept its obvious successes, and not just the financial ones. People enjoy its services and clamor to use them. Amazon gains roughly half of all new online commerce in the United States.[32] Uber[33] and Airbnb[34] have quickly become indispensable to millions in their commutes and travels. Google and Facebook both now serve more than a billion people worldwide, a total that keeps growing.

So where, exactly, is the problem? Perhaps the simplest way to describe it is that the combination of a hacker's arrogance and an entrepreneur's greed has turned a collective enterprise like the Web into something proprietary, where our online profiles, our online relationships, our online posts and Web pages and photographs are routinely exploited for business reasons. Companies now regularly spy on their users as they travel across the Web, and save this information. With the help of "artificially intelligent" algorithms, these companies create profiles to place particularly effective ads before the eyes of their visitors. The public increasingly finds itself at the mercy of the choices of a few dominant tech companies, whose services have become too large and pervasive to ignore.

Facebook has been the most ambitious and most successful in expanding its audience. In 2016, more than half of the United States population visited the site at least once a month, while Facebook shot up to 500 million monthly active users worldwide in 2010, then 1 billion in 2012, with 2 billion expected in 2017.[35] The company is edging toward Zuckerberg's goal of creating "a utility—you know, something that people use in their daily lives to look people up and find information about people."[36] When Zuckerberg was asked if this global utility should be regulated by governments the same way the electric and water companies are, he replied, "In terms of regulation, I mean, we get regulated by users, right?"[37] With that answer, Facebook's chief executive revealed in equal measure the entrepreneur's impatience with regulators who would encroach on his ability to make unlimited profit and the hacker's hubris that he knows how to "regulate" his company better than any government could.

In fact, tech companies believe that through artificial intelligence tools they understand their users' state of mind in a way few other companies can, and far better than any regulator. They can track, measure, and analyze the billions of decisions their users make, and they can detect even the most minor feature that may be turning them off. And rather than wait for problems, these companies can compel their users to express a preference by staging so-called A/B testing, which involves showing groups of users slightly different versions of the site and measuring which group stays longer and is thus happier with the experience. Google famously went so far as to prepare forty-one shades of blue to test which was the best color for displaying links in its Gmail service.

When Douglas Bowman, Google's first visual designer, quit in frustration he cited the "shades of blue" episode as what happens when computer engineers run all aspects of the company, including design: "Reduce each decision to a simple logic problem. Remove all subjectivity and just look at the data. Data in your favor? O.K., launch it. Data shows negative effects? Back to the drawing board. And that data eventually becomes a crutch for every decision, paralyzing the company and preventing it from making any daring design decisions."[38] What Bowman saw as a recipe for uninspired design, Google saw as satisfying the public; after all, the highest-scoring shade of blue was the one that users clicked on the most. The company later claimed that using its superior shade of blue had generated $200 million in extra revenue because more people viewed Google's advertising.[39] As Bowman signed off with a post to his personal blog, he conceded that he had no vocabulary to make his case to Google to change its ways; his appeals to taste and judgment literally did not compute. Thus, the message coming back from Silicon Valley wasn't apologetic to the designer but was insistent that if something can't be measured then it is most likely imaginary, like religion or good design or "truth" or empathy.

With this posture, and in so many other ways, the Know-It-Alls bring to mind a precocious teenager who is sure he knows more than he does, slamming the door to his room and muttering about the "phonies" and "dummies" ruining the world. When Andreessen was well into his forties, he routinely took to Twitter to make cranky, snarky comments about idealists like the French tech minister who was quoted saying that she thought her country could take on Silicon Valley, "but without all that horrible inequality." His sarcastic putdown to her (and presumably this author as well) was, "Capitalism, without all that messy capitalism! :-)."[40] Andreessen has also been quoted in the press making misanthropic comments like "I really don't like people," or expressing a preference for lawn mowers because "your lawn mower never argues with you."[41] His venture capital partner, Ben Horowitz, is similarly bold, writing a harsh business-advice book called *The Hard Thing About Hard Things*, which is peppered with gangsta rap lyrics. Introducing a brief chapter on corporate culture, Horowitz includes a quote from the rapper Trinidad James: "This ain't for no fuck niggas, if you a real nigga then fuck with me."[42]

There are other adolescent obsessions, too, beyond the urge to shock the grownups. Peter Thiel believes that science can banish death, if only we considered it a priority. Elon Musk fears that robots will enslave humans.[43] And just about every Know-It-All holds dear the fantasy and science fiction stories that sustained them during their youth—Thiel's current company, Palantir, which uses sophisticated filtering algorithms to help companies and governments track members of the public, is named after an all-seeing stone from J.R.R. Tolkien's *Lord of the Rings* trilogy. Andreessen, who grew up unhappily in rural Wisconsin, recalled the rare joy of watching the original *Star Wars* in an unheated local theater, while Zuckerberg, who had a *Star Wars*–themed bar mitzvah, recently wrote an enthusiastic comment on the official Star Wars Facebook page about the trailer for Disney's reboot of the franchise—"This looks amazing. I love Star Wars." Disney's social media team responded immediately: "We know."[44]

As all this Star Wars talk demonstrates, the Know-It-Alls can appear to be friendly nerds, akin to the eccentric scientists on the sitcom *The Big Bang Theory*. But Andreessen, Thiel, and Zuckerberg are not scientists; they are civic and economic leaders whose ideas and wealth are too influential to trivialize. It's one thing for an awkward programmer to have a hard time speaking with women—like Raj on *Big Bang*, who in early seasons needed to be drunk to hold a conversation with a member of the opposite sex—and another for Silicon Valley companies to vastly underrepresent women, African Americans, and Hispanics. In fact, there has been remarkable demographic consistency among the employees of companies like Google, Facebook, and Uber, with a tech workforce that is anywhere from 15 percent to 20 percent women, 1 to 2 percent African American, and 2 to 3 percent Hispanic.[45]

Yet the most purely adolescent quality of the Know-It-Alls may well be their seeming glee at the destruction that can be laid at their feet. The Stanford literature professor Robert Pogue Harrison reminds us just how extraordinary such an attitude is. "Our silicon age, which sees no glory in maintenance, but only in transformation and disruption, makes it extremely difficult for us to imagine how, in past eras, those who would change the world were viewed with suspicion and dread," he writes. "If you loved the world; if you considered it your mortal home; if you were aware of how much effort and foresight it had cost your forebears to secure its foundations, build its institutions, and shape its culture; if you saw the world as the place of your secular afterlife, then you had good reasons to impute sinister tendencies to those who would tamper with its configuration or render it alien to you."[46] Instead, we have revered these social tamperers and await with interest for each new manifesto they issue with titles like "Building Global Community" or "What Happened to the Future?"

Back in 2009, Andreessen described the epic battles to come between disruptive companies and traditional businesses. The economist Joseph Schumpeter, famous proponent of creative destruction, "would be proud," Andreessen wrote, as he warned that "companies in every industry need to assume that a software revolution is coming." He rattled off a litany of the old and inefficient organizations that have been toppled or are under siege by new and nimble companies—Blockbuster video rental, Borders bookstores, Kodak film, the United States Postal Service. "Health care and education, in my view, are next up," he predicted in his piece, which is titled "Why Software Is Eating the World," or, if you prefer, "How Software Serves Man: A Cookbook."[47]

1. JOHN McCARTHY

"Solving today's problems tomorrow"

"What do judges know that we cannot eventually tell a computer?" John McCarthy asked himself with a rhetorical flourish in a debate at Stanford about the limits, if any, to artificial intelligence research. In 1973, the answer was obvious to McCarthy: "Nothing."[1] The leader of the university's highly regarded artificial intelligence lab, McCarthy appeared part mad scientist, part radical intellectual, with the horn-rimmed glasses and pocket protector of an engineer and the bushy hair and rough beard of a firebrand. McCarthy's opposite that day was Joseph Weizenbaum, a dapper, pipe-smoking MIT computer science professor, who by the 1970s had come to challenge everything McCarthy stood for. Where McCarthy believed nothing human was beyond the capability of machines when properly instructed, Weizenbaum insisted that some tasks—like sitting in judgment of the accused or giving counsel to those in distress—could only be entrusted to people. Even to consider otherwise was to commit a "monstrous obscenity."[2] A Jewish refugee from Nazi Germany at age thirteen, Weizenbaum was ever on the watch for those whom he suspected didn't believe that all human life was sacred, whether because of a commitment to racial superiority or a conviction that nothing innate separated people from machines.[3]

To Weizenbaum, McCarthy's ease in casting aside the ethical concerns of others was the clearest sign yet that elite AI scientists, whom Weizenbaum called the "artificial intelligentsia," had lost their way. They would sacrifice anything for the cause, even their own humanity. Case in point were the young computer experts—the hackers—whom McCarthy had nurtured in his labs, first at MIT and later at Stanford. "They work until they nearly drop, twenty, thirty hours at a time," Weizenbaum wrote in *Computer Power and Human Reason*, his anti-AI manifesto published in 1976. "Their food, if they arrange it, is brought to them: coffee, Cokes, sandwiches. If possible, they sleep on cots near the computer. But only for a few hours—then back to the console or the printouts. Their rumpled clothes, their unwashed and unshaven faces and their uncombed hair all testify that they are oblivious to their bodies and to the world in which they move. They exist, at least when so engaged, only through and for the computers. These are computer bums, compulsive programmers."[4]

Naturally, this assessment of the stars of his lab struck McCarthy as unfair, but that last slur—bum!—from a fellow computer scientist really stung. First, McCarthy knew that the hackers' enthusiasm, even compulsion, was crucial to running a successful lab, something Weizenbaum apparently didn't need to consider now that he was more interested in ethics than research. "We professors of computer science sometimes lose our ability to write actual computer programs through lack of practice and envy younger people who can spend full time in the laboratory," McCarthy explained to Weizenbaum in a review of *Computer Power*, which he titled "An Unreasonable Book." "The phenomenon is well known in other sciences and in other human activities."[5] But more, McCarthy saw critics like Weizenbaum as lacking the scientist's relentless drive to understand the world, no matter where that drive may take him, perhaps even to programming computers to think like judges. "The term 'appropriate science' suggests that there is 'inappropriate science,'" McCarthy said at a later debate. "If there is 'inappropriate science,' then there is 'appropriate ignorance' and I don't believe in it."[6] His young "computer bums" had nothing to apologize for.

McCarthy came easily to this take-no-prisoners debating style on behalf of science and research—ideological combat was in his blood. As a teenager, John McCarthy wasn't only a mathematics prodigy—skipping two grades at Belmont High School in Los Angeles and teaching himself advanced calculus from college textbooks he bought second-hand—he was also a young Marxist foot soldier, a full member of the Communist Party at the age of seventeen, skipping ahead there, too.[7]

John McCarthy was born back east to parents who represented two prominent streams within American radicalism—the Jewish and the Irish. His mother, Ida Glatt, was a Lithuanian Jewish immigrant who grew up in Baltimore and managed to attend the prestigious women's school Goucher College, probably with the assistance of the city's wealthier German Jewish community. His father, Jack, an Irish Catholic immigrant, hoped to avoid deportation for his political activities by claiming to be a native Californian whose birth certificate went missing in the San Francisco earthquake. Years before Ida and Jack ever met, each had already led popular protests: Ida was at the head of Goucher students marching for women's suffrage, and Jack was urging stevedores in Boston to stop loading a British ship in solidarity with a hunger strike of Terence MacSwiney, a jailed Irish Republican politician.[8]

Though Jack never made it past the fourth grade, his son John remembered his literary temperament, whether that meant quoting Browning and Kipling or reciting poems and lyrics about the Irish cause.[9] Ida was an accomplished student and an idealistic champion for the poor and oppressed. She graduated from Goucher in 1917 with a degree in political economy and went to the University of Chicago to do graduate work that quickly spilled into labor organizing. Due to her refined education and her sex, Ida was brought under the umbrella of the Women's Trade Union League, an organization identified with the wealthy progressive women who helped fund it—the so-called mink brigade. The league's motto connected workers' rights to justice for women and families: "The eight hour day, a living wage, to guard the home."[10]

A few years later, Ida and Jack were introduced in Boston—she hired him to build a set of bookshelves, according to family lore—and Ida was fully radicalized. In addition to his carpentry, Jack ran organizing drives for fishermen and dry-cleaning deliverers, trolley workers and longshoremen. Their first child, John, was born there in 1927, though soon the family moved to New York, where the couple worked for the Communist Party newspaper the *Daily Worker*; Ida was a reporter and Jack was a business manager.[11] For the sake of the health of young John, who had a life-threatening sinus condition, they relocated to Los Angeles, then known for its clean, dry air. Ida became a social worker, and Jack continued labor organizing, serving at one point as an aide to Harry Bridges, the radical San Francisco–based longshoremen's union leader who was West Coast director of the CIO during the 1930s and 1940s.

Young John regained his strength and early on showed an interest in math and science. His parents gave him a party-approved volume, the English translation of a children's science book popular in the Soviet Union, *100,000 Whys*, which ingeniously explains biology, chemistry, and physics by looking at how an ordinary household works—the stove, the faucet, the cupboard. The slim volume opens with the observation, "Every day, somebody in your house starts the fire, heats water, boils potatoes. . . ."[12] Later, he would purchase used calculus textbooks.

In an act of "extreme arrogance," John applied to a single college, the nearby California Institute of Technology. His equivalent of a college essay was a single sentence: "I intend to be a professor of mathematics." John was accepted and completed his work there in two and a half years, though graduation was delayed two years because he was suspended twice for refusing to attend mandatory gym classes. One of the suspensions led to a detour to the army, at nearby Fort MacArthur, where McCarthy and other young enlisted men were entrusted with the calculations that determined whether candidates deserved a promotion—"the opportunity for arbitrary action by just us kids was extraordinary." He harbored little anger at the delay in obtaining his diploma, noting that after World War II the army became a rather genial place. "Basic training was relaxing compared to physical education at Caltech,"[13] he recalled. After graduation in 1948, McCarthy spent a year at Caltech as a mathematics graduate student in preparation for becoming a professor. Two events that year propelled McCarthy toward what would become his lifelong quest: to create a thinking computer.

First, before ever setting eyes on a computer, McCarthy studied how to program one. He attended lectures about the Standards Western Automatic Computer, which wouldn't be completed and installed in Los Angeles until two years later. The word *computer* was being transformed during these years. Once, it had been used to describe a person, usually a woman, who carries out complicated calculations. But by 1948, *computer* could also describe a machine that in theory was able to follow instructions to perform any task, given enough time. This framework for a computer was proposed by the great British mathematician Alan Turing and called the "universal Turing machine." By breaking down any action, up to and including thought, as merely a sequence of discrete steps each of which could be achieved by a computer, Turing would give hope to dreamers like McCarthy and convince them that, in McCarthy's words, "AI was best researched by programming computers rather than by building machines."[14]

Turing was so confident that computers could be instructed how to think that he later devised a way of verifying this achievement when it inevitably arrived, through what he called "the imitation game." Turing's contention with his game, which is now more commonly called the "Turing test," was that if a computer could reliably convince a person that he was speaking with another person, whom he could not see, then it should be considered intelligent. The true potency of the test, writes the historian Paul N. Edwards, was its limited, machine friendly definition of intelligence. "Turing did not require that the computer imitate a human voice or mimic facial expressions, gestures, theatrical displays, laughter, or any of the thousands of other ways humans communicate," Edwards writes. "What might be called the intelligence of the body—dance, reflex, perception, the manipulation of objects in space as people solve problems, and so on—drops from view as irrelevant. In the same way, what might be called social intelligence—the collective construction of human realities—does not appear in the picture. Indeed, it was precisely because the body and the social world signify humanness directly that Turing proposed the connection via remote terminals."[15]

Turing's ideas were highly speculative: computers and mathematicians hardly seemed up to the task of creating an artificial intelligence, even under Turing's forgiving definition. Nonetheless, interest was

building. In 1948, McCarthy attended an unprecedented conference at Caltech, "The Hixon Symposium on Cerebral Mechanisms in
Behavior," which included talks like "Why the Mind Is in the Head"
and "Models of the Nervous System."[16] The featured speaker was the
mathematician and computer pioneer John von Neumann, who cited
Turing to explain how a computer could be taught to think like a
human brain. There also were lectures from psychologists who worked
the other way around, finding parallels between the human brain and
a computer. Within this intellectually charged atmosphere, McCarthy committed himself to studying "the use of computers to behave
intelligently."[17]

When McCarthy departed Caltech, and the family home,
for Princeton the next year, he was on course to earn a PhD in
mathematics—there was no such field as artificial intelligence, or even
computer science for that matter. By virtue of von Neumann's presence
nearby at the Institute for Advanced Study, McCarthy was studying at
one of the few hotbeds for these ideas. He soon met Marvin Minsky,
another mathematics graduate student, who became a friend and an
AI fellow traveler. McCarthy also fell in with a circle of mathematicians, including John Forbes Nash, who were devising game theory,
a scheme for modeling human behavior by describing the actions of
imaginary, self-interested individuals bound by clear rules meant to
represent laws or social obligations.

Still on the left politically, McCarthy hadn't accepted game theory's
cynical view of people and society. He recalled Nash fondly, but considered him peculiar. "I guess you could imagine him as though he
were a follower of Ayn Rand," McCarthy said, "explicitly and frankly
egotistical and selfish." He was there when Nash and others in their
circle helped create a game of deceit and betrayal that Nash called
"Fuck Your Buddy"; McCarthy said he came up with a more family-
friendly name for the game, "So Long, Sucker." It stuck. The ruthless
strategy needed to excel in "So Long, Sucker" offended McCarthy, and
he lashed out at Nash one time as the game descended into treachery:
"I remember playing—you have to form alliances and double cross at
the right time. His words toward me at the end were, 'But I don't need
you anymore, John.' He was right, and that was the point of the game,
and I think he won."[18]

Exposed to the rationalist ideas of thinkers like von Neumann, Nash and Minsky, and others, McCarthy was becoming increasingly intellectually independent. He was finally away from his parents—even in the army he had been assigned to a nearby base—and had the freedom to drift from radical politics. McCarthy tells a story of dutifully looking up the local Communist Party cell when he arrived in Princeton and finding that the only active members were a janitor and a gardener. He passed on that opportunity and quietly quit the party a few years later. During the Red Scare led by a different McCarthy (Senator Joseph) he had to lie a couple of times about having been a Communist, "but basically the people that I knew of who were harmed were people who stuck their necks out."[19] McCarthy didn't, and thus began a steady shift to the right.

Toward the end of his life, when his political transformation from left to right was complete, McCarthy wrote an essay trying to explain why otherwise sensible people were attracted to Marxism. One of the attractions he identified—"the example of hard work and self-sacrifice by individual socialists and communists in the trade union movement"—undoubtedly sprung from the committed political lives of his parents, Ida and Jack. Nonetheless, McCarthy rates the Marxist experiment a terrible blight on human history and observes about his own radical upbringing, "An excessive acquaintance with Marxism-Leninism is a sign of a misspent youth."[20]

McCarthy's "misspent youth," however, is what gives narrative coherence to his life, even if it is the kind of narrative O. Henry liked to tell. It goes something like this: Man spends his life shedding the revolutionary ideology of his upbringing and its dream of a utopian society without poverty or oppression to commit instead to a life of scientific reason. The man becomes an internationally recognized computer genius and runs a prestigious lab. Over time, the man's scientific research fuels a new revolutionary ideology that aspires to create a utopian society without poverty or oppression. In other words, if McCarthy believed as a young man that adopting the rational values of science and mathematics would offer refuge from the emotional volatility of politics, he couldn't have been more mistaken. By the end of his life, McCarthy was more political agitator than scientist, even if he continued to see himself as the epitome of a rational being.

After McCarthy completed his PhD at Princeton in pure math, he had little direction in his academic career—mainly he was struck by what he considered the shallowness of his own mathematical research, especially when compared with the depth of the work of Nash and the others in his circle. McCarthy had original ideas he would speculate on, but he wondered if that was enough for the highest levels of mathematics. McCarthy was hired by Stanford as an assistant professor and then quickly let go—"Stanford decided they'd keep two out of their three acting assistant professors, and I was the third."[21]

McCarthy continued his academic career at Dartmouth, where in 1955 he began planning a summer workshop on thinking computers. There wasn't yet a term describing this type of research. "I had to call it something, so I called it 'artificial intelligence,'" he recalled. An earlier name for the topic, *automata studies*, came from the word for self-operating machines, but didn't describe the goal nearly as breathlessly as *artificial intelligence* did.[22] The Dartmouth summer workshop matched up McCarthy and Minsky with famous names in computing, the information theorist Claude Shannon, who had left Bell Labs for MIT, and Nathaniel Rochester, an IBM researcher, though perhaps the most famous name of them all, von Neumann, would not be there: by the summer of 1956, he was too sick to attend. The youthful arrogance of Minsky and McCarthy was on bright display throughout the proposal to the Rockefeller Foundation asking for $13,500 to cover stipends, travel expenses, and the like, including this succinct description of their core belief: "Every aspect of learning or any other feature of intelligence can in principle be so precisely described that a machine can be made to simulate it."[23]

Buoyed by the success of the conference, which now has its own commemorative plaque on the Dartmouth campus, McCarthy maneuvered himself to MIT. First, he persuaded the head of the mathematics department at Dartmouth, John Kemeny, to arrange a fellowship, which McCarthy elected to take at MIT, and then, "I double-crossed him by not returning to Dartmouth, but staying at MIT," he recalled. By 1958, Minsky had arrived at MIT, too; the next year, the two were running the artificial intelligence project there. MIT was so flush with government money in those years that administrators offered the newly arrived junior professors the funds to support six mathematics graduate students, no questions asked.[24]

McCarthy and Minsky's graduate students were put to work applying their training in mathematical logic to computers; they were asked to find ways to represent the outside environment, as well as a brain's stored knowledge and thought processes, by means of a series of clearly defined statements. In this sense, the early AI project was conceived as a reverse-engineering problem. How to build an intelligent computer? Well, first you "open up" a person and study in detail what makes him tick. For example, Minsky and his graduate students would ask probing questions of children—so probing that Margaret Hamilton, an MIT graduate student at the time and later a famous software engineer, recalls Minsky's team making her three-year-old daughter cry during an experiment that had a researcher read a computer's critical comments back to her. "That's how they talked back then; they thought computers were going to take over the world then," Hamilton said.[25] In one unpublished paper, called "The Well-Designed Child," McCarthy tried to detail what we know and don't know about the tools for reasoning that are born to human babies. A short section, titled "Unethical Experiment," shows just how curious he was: "Imagine arranging that all a baby ever sees is a plan of a two-dimensional room and all his actions move around in the room. Maybe the experiment can be modified to be safe and still be informative."[26]

From what AI researchers deduced about people, by experiment or intuiting, they devised computer code to imitate the process step by step, algorithm by algorithm. In an essay for *Scientific American* in 1966, McCarthy made the same confident assertion he had first expressed as part of the Dartmouth conference, that nothing, in theory, separated a computer from a person. "The computer accepts information from its environment through its input devices; it combines this information, according to the rules of the program stored in its memory, with information that is also stored in its memory, and it sends information back to its environment through its output devices," he wrote, which was just the same as people. "The human brain also accepts inputs of information, combines it with information stored somehow within and returns outputs of information to its environment."[27] McCarthy's every instinct was to demystify the process of human thinking and intelligence. Behavior could be explained by "the principle of rationality"—setting a goal (not necessarily a rational goal) and then coming up with a plan to achieve it.[28]

The tricky thing for a computer to replicate was ordinary common sense, not differential calculus, McCarthy concluded. By 1968, a robot in McCarthy's lab had an arm with refined touch, yet it still could not tie a pair of shoes. Maddening. "I have observed two small children learn how to do this, and I don't understand how they learn how to do it," he said. "The difficulty in this case is not so much in getting the sense itself but programming what to do with it."[29] Thus, McCarthy spent a lot of time trying to understand precisely how people got through daily life. He was forever challenging his intelligent machines with "mundane and seemingly trivial tasks, such as constructing a plan to get to the airport," or reading and comprehending a brief crime report in the *New York Times*.[30]

In his pursuit of a thinking machine, McCarthy was making a bunch of dubious leaps, as he later came to acknowledge. To deconstruct a human being necessarily meant considering him in isolation, just as the designers of an intelligent machine would be considering it in isolation. Under this approach to intelligence there would be no pausing to consider the relevance of factors external to the brain, such as the body, social ties, family ties, or companionship. When people spoke about the mystery of the human mind, McCarthy would scoff. Could a machine have "free will"? A colleague recalled his answer, which sought to remove the sanctity of the term: "According to McCarthy, free will involves considering different courses of action and having the ability to choose among them. He summarized his position about human free will by quoting his daughter, Sarah, who said at age four, 'I can, but I won't.'"[31]

In the spring of 1959, four years after he finally learned to program properly, McCarthy taught the first freshman class in computers at MIT, turning on a generation of aspiring programmers who followed him to the artificial intelligence lab.[32] The curriculum was brand-new, and the term *computer science* had only just surfaced. One of its first uses was in a report commissioned by Stanford around 1956 as it considered whether the study of computers could be an academic discipline. Stanford's provost, Frederick Terman, was always thinking of ways to raise the university's profile and add to its resources, and there was a tempting offer from IBM that it would donate state-of-the-art equipment, including its pathbreaking 650 mainframe, to any university that offered classes on scientific and business computing. Stanford asked a local computer expert, Louis Fein, to study the topic and propose an academically rigorous curriculum in the "science" of computers.

The university's mathematics faculty was skeptical. One professor told Fein that he thought computing was like plumbing, a useful tool to be applied to other projects. "We don't have plumbers studying within the university," he told Fein. "What does computing have to do with intellect?" But Fein was inclined to flip the question around in his head. "Why is it that economics and geology is in a university," he'd ask himself, "and why is it that plumbing isn't?"[33] For his report, "The Role of the University in Computers, Data Processing, and Related Fields,"[34] Fein interviewed the nation's computer experts—a total in the "tens" at the time, he recalled—including McCarthy, when he was still at Dartmouth, and Weizenbaum, when he was a researcher at Wayne State University. Fein recommended that Stanford move forward in creating a separate department for computer science, and the university, as a first step, introduced a division within the mathematics department.

Stanford's administrators ignored an unstated conclusion in the report: namely, that Fein should be brought in to lead Stanford's new computer science department. Terman's deputy made it clear to Fein that Stanford would only be recruiting big names—"what we need is to get a von Neumann out here and then things will go well," he was told.[35] McCarthy might fit the bill, however. In 1962, Stanford offered him a big promotion—immediate tenure as a full professor in mathematics—as a prelude to joining a newly minted computer science department. Yet the quick return and advancement of McCarthy at Stanford after initial failure would give ammunition to those mathematicians who argued that computer work must not be rigorous. Wasn't this big shot McCarthy only just passed over as a junior professor?

Intelligence was the coin of the realm in those years in a preview of how Silicon Valley would operate—that is, everyone was intent on identifying who was most brilliant and rewarding him, all the while looking for the magic formula to mass-produce intelligence in the computer lab. Not even McCarthy was above reproach. No one ever is, really. In 2014, the Web site Quora began a discussion, "How did Mark Zuckerberg become a programming prodigy?" Coders were eager to burst that bubble and explain that Zuckerberg, despite his success, was no prodigy. "The application he wrote was not unique, and not all that well-made," one Quora contributor explained.[36]

Any doubts about McCarthy's brilliance, based on his early failure as a mathematics professor, weren't fair, of course. McCarthy had found his academic purpose within computer science. While at MIT, he had invented a powerful programming language called Lisp, which allowed complicated instructions to be given to the computer with relatively simple commands and is still used today. Paul Graham, the cofounder of the company Y Combinator, which encourages and invests in start-ups, considers McCarthy a hero programmer. "In 1958 there seem to have been two ways of thinking about programming. Some people thought of it as math . . . others thought of it as a way to get things done, and designed languages all too influenced by the technology of the day. McCarthy alone bridged the gap. He designed a language that was math. But designed is not really the word; discovered is more like it," Graham writes in appreciation.[37] Another academic achievement of McCarthy's at MIT was the AI lab itself, which was filling up with eager young programmers who lived and breathed computers.

McCarthy had recognized early on that for the lab to hum with breakthroughs—and for such breakthroughs to build on each other—there had to be more computer time to spread around. The old system had been too limiting and too forbidding: hulking IBM mainframes guarded by specially trained operators who behaved as priests at the ancient temple. Students and the staff skirmished frequently as the hackers tried to trick the IBM guardians into leaving their post.[38] The balance of power began to shift toward the hackers by the end of the 1950s with the arrival of the TX-0, a hand-me-down computer from a military electronics laboratory affiliated with MIT. The TX-0 did not need special operators. In 1961, it was joined by a donated prototype of the PDP-1, the first minicomputer from Digital Equipment Corporation, a Boston-area start-up founded by former MIT scientists who had worked on the TX-0. McCarthy proposed "time-sharing"[39] as a way of replicating the flexibility of these new computers on the IBM mainframe while removing the bottleneck for computer time by replacing one-at-a-time use of a computer with a system that allowed as many as twenty-four people to connect individually with a computer. Eventually each person would have his own monitor and keyboard.[40]

Indeed, once the students interacted with a computer directly the seduction was complete. Young people camped out at McCarthy and Minsky's lab, waiting often into the wee hours of the night for a time slot to open up. When they were dissatisfied with the performance of a vital piece of code that came with the PDP-1, the assembler, they asked the lab for the assignment of writing a better one and were given a weekend to do it. The technology writer Steven Levy recounts the "programming orgy" that ensued: "Six hackers worked around 250 man-hours that weekend, writing code, debugging and washing down take-out Chinese food with massive quantities of Coca-Cola." Digital asked the hackers for a copy of the new assembler to offer to other owners of the PDP-1. They eagerly complied, Levy writes, and "the question of royalties never came up. . . . Wasn't software more like a gift to the world, something that was reward in itself?"[41]

The ecstasy from directly interacting with the computer—not the chance for profits—was the basis of what became the Hacker Ethic, a set of principles these young programmers lived by as they pursued a personal relationship with computing. This first principle was called the "hands-on imperative," the belief that there could be no substitute for direct control over a computer, its inner workings, its operating system, its software. The deeper you went, the better—whether that meant manipulating the so-called machine language that the computer used at its core, or opening up the box to solder new pathways in the hardware that constituted the computer's brain. The access not only had to be direct, but casual, too. "What the user wants," McCarthy wrote at the time, "is a computer that he can have continuously at his beck and call for long periods of time."[42]

If this all sounds sort of sexual—or like an old-fashioned marriage—well, you aren't the first to notice. McCarthy's lab may as well have had a No Girls sign outside. Levy described this first generation of hackers at MIT as being locked in "bachelor mode," with hacking replacing romantic relationships: "It was easy to fall into—for one thing, many of the hackers were loners to begin with, socially uncomfortable. It was the predictability and controllability of a computer system—as opposed to the hopelessly random problems in a human relationship—which made hacking particularly attractive."[43] Margaret Hamilton, the MIT graduate student who later led the software team for the Apollo space mission, would occasionally visit the AI lab and clearly had the chops to be one of the "hackers." She says she had a similar mischievous approach to computing and "understood these kids and their excitement." Even so, she kept a healthy distance. The age gap may have been small, but Hamilton must have seemed a generation older. In her early twenties, she was already married with a child; her computer talents were focused on a practical problem, helping to interpret research within MIT's meteorology department. Hamilton remembers wondering how it was that the AI lab could always be filled with undergraduates. When she was studying math as an undergraduate at Earlham College, she didn't remember having so much free time: "These kids weren't worried about bad marks and satisfying their professors?"[44]

Of course, hackers didn't care about grades. Computers had changed everything. There were other principles to the Hacker Ethic, as explicated by Levy—including "mistrust authority," "all information should be free," and "hackers should be judged by their hacking, not bogus criteria such as degrees, age, race, or position"—but each was really a reframing of the hands-on imperative. The goal was to get close to a computer, and anyone or anything that stood in your way was the enemy. In the competition for computer time, for example, the higher up the academic ladder you were, the greater access you had, whether you were adept or not. That isn't right. Programming skill, not age or academic degrees, is all that should matter in deciding who gets his hands on a computer. Later, there were the businesses that insisted on charging for the programs necessary for you to operate your computer. No way. Information should be free and lack of financial resources shouldn't be an impediment to programming.

These young men were fanatically devoted to computers, which they valued for being more reliable, helpful, and amusing than people. The hackers dreamed of spreading their joy. "If *everyone* could interact with computers with the same innocent, productive, creative impulse that hackers did, the Hacker Ethic might spread through society like a benevolent ripple," Levy wrote in 1984, "and computers would, indeed, change the world for the better."[45]

By 1962, the thirty-five-year-old McCarthy had already made a series of important professional contributions, arguably the most important of his career: the Lisp programming language, time-sharing, an early framework for instructing a computer to play chess, a quintessentially intellectual activity once considered safely beyond machines. Nonetheless, as Stanford quickly became aware, McCarthy was ripe for the picking. He was annoyed that MIT was dragging its feet in implementing his beloved time-sharing idea; an offer of full tenure there would be a few years off, if it came at all. The cold Cambridge winters weren't helping matters, either. McCarthy accepted Stanford's offer.

McCarthy was a big get for Stanford, and not just because he was coming from the pinnacle, MIT. McCarthy was an academic rock star, the kind of professor students wanted to work with. He was indulgent of his students; his field, artificial intelligence, was on the cutting edge. By contrast, George Forsythe, the first computer science professor at Stanford and the first chairman of the department, had a wonky specialty, numerical analysis. At least he had a sense of humor about it. "In the past 15 years," he wrote in 1967, "many numerical analysts have progressed from being queer people in mathematics departments to being queer people in computer science departments!"[46] AI research and programming languages and systems, on the other hand, would be where the growth in computer science would occur because they "seem more exciting, important and solvable at this particular stage of computer science."[47] McCarthy was set up in a remote building that had been abandoned by a utility company, where he went to work re-creating the MIT lab as the Stanford Artificial Intelligence Lab (SAIL).

"What you get when you come out to Stanford is a hunting license as far as money is concerned," McCarthy recalled, and he already had well-established, MIT-based connections to DARPA, the defense department's research group. "I am not a money raiser," McCarthy said. "I'm not the sort of person who calls somebody in Washington and makes an appointment to see him and says, look here, look at these great things, why don't you support it? As long as you could get money by so to speak just sending away for it, I was O.K., but when it took salesmanship and so forth then I was out of it."[48] These government funds came with barely any strings attached, and no supervision to speak of: "I was the only investigator with a perfect record," he liked to say. "I never handed in a quarterly progress report."[49] Because of this benign neglect, McCarthy was able to use money assigned to artificial intelligence research to support a series of important improvements in how computers worked, including individual display terminals on every desk, computer typesetting and publishing, speech recognition software, music synthesizers, and self-driving cars. By 1971, SAIL had seventy-five staff members, including twenty-seven graduate students pursuing an advanced degree, who represented a range of fields: mathematics, computer science, music, medicine, engineering, and psychiatry.[50]

The transfer from MIT to Stanford led to some California-esque adjustments. To start, there was a sauna and a volleyball court.[51] And while still members of a boys' club, SAIL's hackers were more likely to take notice of the opposite sex. A *Life* magazine profile from 1970 helped establish Stanford's reputation as the home of a bunch of horny young scientists, so different from asexual MIT. The article quoted an unnamed member of the team programming a computer psychiatrist and "patient" as saying they were expanding the range of possible reactions a computer can experience: "So far, we have not achieved complete orgasm."[52] The lab's administrators insisted to the student newspaper that the quote was made up, but a reputation was taking shape.[53] In one notorious episode a year later, some computer science students shot a pornographic film in the lab for their abnormal psychology class, recruiting the actress through an ad in the student newspaper seeking an "uninhibited girl."[54] The film was about a woman with a sexual attraction to a computer—for these purposes, one of the experimental fingers attached to a computer used in robotics research proved an especially useful prop. The entire lab was able to watch the seduction scene through an in-house camera system connected to the time-sharing terminals dispersed throughout the building.[55]

At this point in the history of AI, researchers were intrigued by the idea of a computer having sex with a woman, which was central to the plot of *Demon Seed*, a 1970s horror film. Makes sense: there was still confidence (or fear) that AI researchers would succeed in making independent thinking machines to rival or surpass humans. The male scientists identified with the powerful computer they were bringing to life. When the goals for AI had diminished, and computers would only imitate thought rather than embody it, the computers became feminized, sex toys for men in movies like *Weird Science*, *Ex Machina*, and *Her*.

Despite the sexual hijinks and the California sun, the Stanford lab was otherwise quite similar to the MIT lab in its isolation and inward-looking perspective. Located five miles outside of campus, SAIL provided built-in sleeping quarters in the attic for those hackers who wouldn't or couldn't leave; the signs on the doors were written in the elvish script invented for *The Lord of the Rings*. One visitor from *Rolling Stone* described the scene in 1972: "Long hallways and cubicles and large windowless rooms, brutal fluorescent light, enormous machines humming and clattering, robots on wheels, scurrying arcane technicians. And, also, posters and announcements against the Vietnam War and Richard Nixon, computer print-out photos of girlfriends, a hallway-long banner SOLVING TODAY'S PROBLEMS TOMORROW."[56]

During these rocking, carefree years, McCarthy and his team were forced to recognize just how flawed the original artificial intelligence project was. They were stuck in a research quagmire, as McCarthy freely admitted to a reporter for the student newspaper: "There is no basis for a statement that we will have machines as intelligent as people in 3 years, or 15 years, or 50 years, or any definite time. Fundamental questions have yet to be solved, and even to be formulated. Once this is done—and it might happen quickly or not for a long time—it might be possible to predict a schedule of development."[57] By accepting this new reality, McCarthy freed himself to write a series of far-sighted papers on how computers could improve life without achieving true artificial intelligence.

In one such paper, from 1970, McCarthy laid out a vision of Americans' accessing powerful central computers using "home information terminals"—the time-sharing model of computer access brought to the general public.[58] The paper describes quite accurately the typical

American's networked life to come, with email, texting, blogs, phone service, TV, and books all digitally available through what is the equivalent of the digital "cloud" we rely on today. He was so proud of that paper that he republished it with footnotes in 2000, assessing his predictions. Over and over, the footnote is nothing more than: "All this has happened." He allowed, however, "there were several ways things happened differently."[59] For example, his interests were plainly more intellectual than most people's, so while he emphasized how the public would be able to access vast digital libraries, "I didn't mention games—it wasn't part of my grand vision, so to speak."[60]

McCarthy's grand vision for domestic computing was notable for being anti-commercial. He predicted that greater access to information would promote intellectual competition, while "advertising, in the sense of something that can force itself on the attention of a reader, will disappear because it will be too easy to read via a program that screens out undesirable material." With such low entry costs for publishing, "Even a high school student could compete with the *New Yorker* if he could write well enough and if word of mouth and mention by reviewers brought him to public attention." The only threat McCarthy could see to the beautiful system he was conjuring were monopolists, who would try to control access to the network, the material available, and the programs that ran there. McCarthy suspected that the ability of any individual programmer to create a new service would be a check on the concentration of digital power, but he agreed, "One can worry that the system might develop commercially in some way that would prevent that."[61] As, indeed, it has.

Just as AI research was on the wane, McCarthy's lab became a target of radical students, who cited SAIL's reliance on defense department funds to link the work there, however indirectly, to the war in Vietnam. In 1970, a protester threw an improvised bomb into the lab's building—fortunately, it landed in an empty room and was quickly doused by sprinklers. Lab members briefly set up a patrol system to protect their building, while McCarthy responded by taking the fight to the enemy.[62] When anti-war protestors interrupted a Stanford faculty meeting in 1972 and wouldn't leave, the university president adjourned the meeting. McCarthy remained, however, telling the protestors, "The majority of the university takes our position, so go

to hell." When they responded by accusing him and his lab of helping carry out genocide in Vietnam, McCarthy responded: "We are
not involved in genocide. It is people like you who start genocide."[63]
Six years later, McCarthy debated, and had to be physically separated from, the popular Stanford biology professor Paul Ehrlich, who
warned that humanity was destroying the environment through over-
population.[64] McCarthy's disdain for Ehrlich could be summarized in
his observation that "doomsaying is popular and wins prizes regardless
of how wrong its prophecies turn out to be."[65]

In Joseph Weizenbaum, however, McCarthy found a more persistent and formidable critic, one who spoke the same technical language.
The two wrote stinging reviews of each other's work: McCarthy would
berate Weizenbaum for foggy thinking that paved the way for authori-
tarian control of science; Weizenbaum, the AI apostate, insisted on
bringing morality into the equation. Weizenbaum also questioned
the self-proclaimed brilliance of his peers. He, for one, chose to study
mathematics because, "Of all the things that one could study, mathematics seemed by far the easiest. Mathematics is a game. It is entirely
abstract. Hidden behind that recognition that mathematics is the easiest is the corresponding recognition that real life is the hardest."[66]

Weizenbaum was born in Berlin in 1923, the second son of a furrier,
Jechiel, and his wife, Henrietta. After the anti-Semitic laws of 1935,
Weizenbaum's family made their way to the United States by way of
Bremen. His studies were interrupted by service in the U.S. Army Air
Corps during World War II,[67] but by the early 1950s, he was a graduate student in mathematics at Wayne State University, where a professor had decided to build a computer from scratch. Weizenbaum was
part of the group that assembled and programmed that computer, even
soldering the components. Based on that experience, Weizenbaum
in 1952 was asked to join the team that produced the huge check-
processing computer for Bank of America, the first for any bank.[68]

All that mattered to Weizenbaum at the time were the technical
challenges that lay ahead—how to teach a computer to read checks
that were crumpled, or of different sizes. "It was a whale of a lot of fun
attacking those hard problems, and it never occurred to me at the time
that I was cooperating in a technological venture which had certain

social side effects which I might come to regret," he said. "I was totally wrapped up in my identity as a professional, and besides, it was just too much fun."[69] In other words, when Weizenbaum, in *Computer Power and Human Reason*, described the intoxication that comes from being a compulsive programmer he was sharing firsthand experiences. "The computer programmer," he wrote, "is a creator of universes for which he alone is the lawgiver. . . . No playwright, no stage director, no emperor, however powerful, has ever exercised such absolute authority to arrange a stage or a field of battle and to command such unswervingly dutiful actors or troops."[70]

After Weizenbaum arrived at MIT in 1963 as an associate professor in electrical engineering, his familiarity with computers led him toward artificial intelligence and an intriguing project he called Eliza (named for Eliza Doolittle of *Pygmalion*). Eliza was programmed to ask a user probing, if vague, questions in the manner of a therapist. A sample conversation between a "patient," played by Weizenbaum, and Eliza, circa 1966, is quite revealing, proving how adept Eliza was at pushing all the right (human) buttons.[71]

PATIENT. Men are all alike.

ELIZA. In what way?

PATIENT. They're always bugging us about something or other.

ELIZA. Can you think of a specific example?

PATIENT. Well, my boyfriend made me come here.

ELIZA. Your boyfriend made you come here?

PATIENT. He says I'm depressed much of the time.

ELIZA. I am sorry to hear you are depressed.

PATIENT. It's true. I am unhappy.

ELIZA. Do you think coming here will help you not to be unhappy?

PATIENT. I need some help, that much seems certain.

ELIZA. What would it mean to you if you got some help?

PATIENT. Perhaps I could learn to get along with my mother.

ELIZA. Tell me more about your family.

PATIENT. My mother takes care of me.

ELIZA. Who else in your family takes care of you?

PATIENT. My father.

ELIZA. Your father.

PATIENT. You are like my father in some ways.

ELIZA. What resemblance do you see?

PATIENT. You are not very aggressive but I think you don't want me to notice that.

ELIZA. What makes you think I am not very aggressive?

PATIENT. You don't argue with me.

ELIZA. Why do you think I don't argue with you?

PATIENT. You are afraid of me.

ELIZA. Does it please you to believe I am afraid of you?

Eliza was meant to explore an obscure concept, "knowledge representation," but became a turning point for Weizenbaum's career; he and his software were a media phenomenon. TV cameras arrived at the lab. In 1968, the *New York Times* headlined a report on Eliza, "Computer Is Being Taught to Understand English." The *Times* reported that Weizenbaum's secretary apparently felt so connected to Eliza that she was offended when he casually presumed he could eavesdrop on her conversation. After typing a few sentences, she turned to him to say, "Would you mind leaving the room, please?" Weizenbaum was taken aback.[72]

When confronted with his power to manipulate people with relatively simple coding, and then to have access to their most personal thoughts, Weizenbaum retreated in horror. He began to ask hard questions of himself and colleagues. The artificial intelligence project, he concluded, was a fraud that played on the trusting instincts of people. Weizenbaum took language quite seriously, informed in part by how the Nazis had abused it, and in that light he concluded that Eliza was lying and misleading the very people it was supposed to be helping. A comment like "I am sorry to hear you are depressed" wasn't true and should appall anyone who hears it. A computer can't feel and Eliza isn't "sorry" about anything a patient said.

As we'll see in the pages that follow, the Know-It-Alls have moved boldly ahead where Weizenbaum retreated, eager to wield the almost hypnotic power computers have over their users. Facebook and Google don't flinch when their users unthinkingly reveal so much about themselves. Instead, they embrace the new reality, applying programming power to do amazing—and amazingly profitable—things with all the information they collect. If anything, they study how to make their users comfortable with sharing even more with the computer. Weizenbaum becomes an example of the path not taken in the development of computers, similar, as we'll see, to Tim Berners-Lee, the inventor of the World Wide Web. Weizenbaum and Berners-Lee each advocated stepping back from what was possible for moral reasons. And each was swept away by the potent combination of the hacker's arrogance and the entrepreneur's greed.

Weizenbaum kept up the fight, however, spending the remainder of his life trying to police the boundary between humans and machines, challenging the belief, so central to AI, that people themselves were nothing more than glorified computers and "all human knowledge is reducible to computable form."[73] This denial of what is special about being human was the great sin of the "artificial intelligentsia," according to Weizenbaum. In later debates with AI theorists, he was accused of being prejudiced in favor of living creatures, of being "a carbon-based chauvinist."

One bizarre manifestation of this charge that Weizenbaum favored "life" over thinking machines (should any arrive) came during a discussion with Daniel Dennett and Marvin Minsky at Harvard University. Weizenbaum recalled that "Dennett pointed out to me in just so many words: 'If someone said the things that you say about life about the white race in your presence, you would accuse him of being a racist. Don't you see that you are a kind of racist?'"[74] Weizenbaum couldn't help but consider this accusation as the abdication of being "part of the human family," even if he suspected that the drive behind artificial intelligence researchers like McCarthy, almost all of whom are men, was all too human. "Women are said to have a penis envy, but we are discovering that men have a uterus envy: there is something very creative that women can do and men cannot, that is, give birth to new life," he told an interviewer. "We now see evidence that men are striving to do that. They want to create not only new life but better life, with fewer imperfections. Furthermore, given that it is digital and so on, it will be immortal so they are even better than women."[75]

Weizenbaum never could have imagined how the Know-It-Alls, beginning in the 1990s, managed to amass real-world wealth and power from the imaginary worlds on their screens and the submerged urge to be the bestower of life. His later years were spent in Germany, far removed from the events in Silicon Valley.[76] However, in 2008, a few months before he died in his Berlin apartment at age eighty-five, Weizenbaum did share the stage at the Davos World Economic Forum with Reid Hoffman, the billionaire founder of LinkedIn, and Philip Rosedale, the chief executive of the pathbreaking virtual reality site Second Life. Hoffman explained why LinkedIn was such an important example of "social software": "What happens is this is my expression of myself, and part of what makes it social is it is me establishing my identity. It's me interacting with other people." Weizenbaum shakes his head in disbelief. People were misleading each other about their "identities" via computers much the way his Eliza program misled the people who communicated with it. Speaking in German, which was simultaneously translated, Weizenbaum tried and failed to rouse the crowd. "Nonsense is being spouted. Dangerous nonsenses. . . . You've already said twice, 'it's happening and it will continue'—as if technological progress has become autonomous. As if it weren't created by human beings. Or that people are doing it for all sorts of intentions. The audience is just sitting here, and no one is afraid, or reacting. Things are just happening."[77]

As the 1970s ended, so, too, did McCarthy's independent laboratory. By 1979, DARPA's funding was largely eliminated, and SAIL merged with the Stanford Computer Science Department and relocated on campus. McCarthy the scientist was already in recess, but McCarthy the polemicist had one last great success: he led the fight against censorship of the Internet, and arguably we are still dealing with its consequences today.

The World Wide Web wasn't up and running in early 1989, but computers at universities like Stanford were already connected to each other via the Internet. People would publish, argue, and comment on Usenet "newsgroups"—conversations organized around particular subjects—which were accessible through computers connected to the university's network. The newsgroups reflected the interests of computer users at the time, so if you think today's Internet is skewed toward young men obsessed with science fiction and video games, just imagine what the Internet was like then. Some newsgroups were moderated, but frequently they were open to all; send a message and anyone could read it and write a response. At the time, Stanford's electronic bulletin board system hosted roughly five hundred newsgroups with topics ranging from recipes to computer languages to sexual practices.

The controversy began with a dumb joke about a cheap Jew and a cheap Scotsman on the newsgroup rec.humor.funny.[78] The joke was older than dirt and not very funny: "A Jew and a Scotsman have dinner. At the end of the dinner the Scotsman is heard to say, 'I'll pay.' The newspaper headline next morning says, 'Jewish ventriloquist found dead in alley.'" But it landed at an awkward time. Stanford in the late 1980s was consumed by the identity politics ferment. The university was slowly becoming more diverse, which led the administration to replace required courses on Western civilization with a more inclusive curriculum featuring texts by women and people of color. Similarly, there were demonstrations demanding that Stanford offer greater protection to minorities and women on campus who didn't feel fully welcome.[79] These changes brought a backlash as well, leading Peter Thiel and another undergraduate at the time to start the conservative magazine *Stanford Review* to fight the move toward "multiculturalism," which, Thiel later said, "caused Stanford to resemble less a great university than a Third World country."[80]

With these charged events in the background, Stanford administrators decided to block the rec.humor.funny newsgroup, which included a range of offensive humor, from appearing on the university's computers. As McCarthy never tired of pointing out, no one at Stanford had ever complained about the joke. An MIT graduate student was first to object by contacting the Canadian university that hosted the newsgroup. The university cut its ties, but the man who ran the newsgroup found a new host and wouldn't take down the joke. Stanford administrators learned of the controversy in the winter of 1989 and directed a computer science graduate student to block the entire newsgroup. The administrators, for once, were trying to get ahead of events, but they were also presuming to get between the university's hackers and their computers. The graduate student whose expertise was needed to carry out the online censorship immediately informed McCarthy.[81]

McCarthy, the computer visionary, saw this seemingly trivial act as a threat to our networked future. "Newsgroups are a new communication medium just as printed books were in the 15th century," he wrote. "I believe they are one step towards universal access through everyone's computer terminal to the whole of world literature." The psychological ease in deleting or blocking controversial material risked making censorship easy to carry out. No book need be taken off the shelves and thrown out, or, god forbid, burned. Would the public even recognize that "setting up an index of prohibited newsgroups is in the same tradition as the Pope's 15th century Index Liber Prohibitorum"?[82] He rallied his peers in the computer science department to fight for a censorship-free Internet.

Throughout this campaign, McCarthy barely acknowledged the racial tensions that had so clearly influenced the university's actions. He once discussed what he saw as the hypersensitivity of minority groups with a professor who approved of the censorship and was amazed to learn that this professor believed that a minority student might not object to a racist joke because of "internalized oppression." McCarthy was suspicious of this appeal to a force that was seemingly beyond an individual's control.[83] The question of racism finally managed to intrude in the internal discussions of the computer science department through William Augustus Brown Jr., an African American medical student at Stanford who was also studying the use of artificial intelligence in treating patients.

Brown was the lone voice among his fellow computer scientists to say he was glad that "For once the university acted with some modicum of maturity."[84] Drawing from his personal experience, Brown described the issue quite differently than McCarthy and the overwhelmingly white male members of the computer science department had. "Whether disguised as free speech or simply stated as racism or sexism, such humor IS hurtful," he wrote to the bulletin board. "It is a University's right and RESPONSIBILITY to minimize such inflammatory correspondence in PUBLIC telecommunications." He saw what was at stake very differently, too. "This is not an issue of free speech; this is an issue of the social responsibility of a University. The University has never proposed that rec.humor's production be halted—it has simply cancelled its subscription to this sometimes offensive service. It obviously does NOT have to cater to every news service, as anyone who tries to find a Black magazine on campus will readily discover."[85]

McCarthy never responded directly, or even indirectly, to Brown, but others in his lab did, offering an early glimpse at how alternative opinions would be shouted down or patronized online. These responses from the Stanford computer science department today might collectively be called "whitesplaining." One graduate student responded to Brown, "I am a white male, and I have never been offended by white male jokes. Either they are so off-base that they are meaningless, or, by having some basis in fact (but being highly exaggerated) they are quite funny. I feel that the ability to laugh at oneself is part of being a mature, comfortable human being."[86] Others suggested that Brown didn't understand his own best interests. "The problem is that censorship costs more than the disease you're trying to cure. If you really believe in the conspiracy, I'm surprised that you want to give 'them' tools to implement their goals," a graduate student wrote.[87]

The reactions against him were so critical that Brown chose a different tack in reply. He opened up to his fellow students about his struggles at Stanford as a black man. "Having received most of my pre-professional training in the Black American educational system, I have a different outlook than most students," Brown wrote. "I certainly didn't expect the kind of close, warm relationships I developed at Hampton University, but I was not prepared for the antagonism. I don't quite know if it's California, or just Stanford, but . . . I don't know how many times I have had the most pompous questions asked of me; how many times a secretary has gone out of her way to make my day miserable. I sincerely doubt any of my instructors even know my name, although I am in the most difficult program in the medical center. Even my colleagues in my lab waited until last month to get the courage to include me in a casual conversation for the first time, although I have been working there five months." He continued: "I don't really mind the isolation—I can still deal, and it gives me PLENTY of time to study. But I really don't like the cruel humor. Once you come down from the high-flying ideals, it boils down to someone insisting on his right to be cruel to someone. That is a right he/she has, but NOT in ALL media."[88]

Needless to say, such displays of raw emotion were not typical of the communication on the computer science department's bulletin board. No one responded directly to what Brown had shared about his struggles as an African American medical student and computer scientist at Stanford; they continued to mock his ideas as poorly thought out and self-defeating. The closest there was to a defense of Brown was the comment of one graduate student who said he didn't agree with the censorship but worried that many of his peers believed that "minority groups complain so much really because they like the attention they get in the media. People rarely consider the complaints and try to understand the complaints from the minority point of view." He ended his email asking, "Do people feel that the environment at Stanford has improved for minority students? Worsened? Who cares?" Based on the lack of response, Who Cares carried the day.

The twenty-five years since haven't eased the pain for Brown, who left Stanford for Howard University Medical School and today is head of vascular surgery at the Naval Medical Center in Portsmouth, Virginia. The attitude at Stanford, he recalled, was elitist and entitled: "If you came from a refined enough background you could say whatever you wanted. Somehow the First Amendment was unlimited and there was no accountability. Any time you wanted to hold anyone accountable it was un-American. But those comments are neither American nor respectful." The lack of engagement from his peers was "very typical," he said. "It was isolationist there, almost hostile. Hostile in a refined way toward anyone who was different. Dismissive. That's my experience. Unfortunately, I see that attitude today, it doesn't matter whether it's Stanford or the alt-right."

What was particularly painful for Brown was that, other than for his skin color, this was his tribe—he was a hacker, too, who taught himself how to manipulate phone tones, who could discuss the elements in the blood of the Vulcans on *Star Trek*. Yet, he recalled, "As a minority, you are in the circle and not in the circle." He never felt comfortable retreating from society and going so deep into computers. The AI movement, he said, was based on the idea that "I'll be a great person because I will have created something better than me. But that misses the whole point of life—the compassion for humanity. That has nothing to do with how you do in school, whether you can program in seven different languages. Compassion for others, that is the most complex problem humanity has ever faced and it is not addressed with technology or science."[89]

No personal testimony posted to the Stanford bulletin board, no matter how gripping, would ever persuade McCarthy to see the issue of censorship as a matter of empathy for the targets of hate speech. To his mind, there was no such thing as inappropriate science or inappropriate speech. Others may disagree, he allowed. "Stanford has a legal right to do what its administration pleases, just as it has a legal right to purge the library or fire tenured faculty for their opinions," he wrote in an email to the computer science department. But he predicted the university would pay a price in loss of respect in the academic world if the authorities were given control over the Internet. McCarthy's hackers didn't respect authority for its own sake, and he was no different—letting the administrators responsible for information technology at the university decide what could be read on the computers there, he contended, was like giving janitors at the library the right to pick the books.[90]

McCarthy's colleagues in computer science innately shared his perspective; the department unanimously opposed removing the rec .humor.funny newsgroup from university computers. The computer science students overwhelmingly backed McCarthy as well, voting in a confidential email poll, 128 to 4.[91] McCarthy was able to win over the entire university by enlisting a powerful metaphor for the digital age. Removing a newsgroup, he explained to those who may not be familiar with them, was like removing a book from the library system because it was offensive. Since *Mein Kampf* was still on the shelves, it was hard to imagine how the decision to remove an anti-Semitic, anti-Scottish joke would make the cut. Either you accepted offensive speech online or you were in favor of burning books. There would be no middle ground permitted, and thus no opportunity to introduce reasonable regulations to ensure civility online, which is our predicament today.

The newsgroup and the dumb joke were restored in a great victory for McCarthy, which took on greater meaning in the years that followed, when the Web brought the Internet to even more parts of the university. Stanford had agreed that "a faculty member or student Web page was his own property, as it were, and not the property of the university," McCarthy told a crowd gathered for the fortieth anniversary of the Stanford Computer Science Department in 2006.[92]

This achievement represented McCarthy's final act in support of the hackers he had helped introduce to the world. He had ensured that their individualistic, anti-authoritarian ideas would be enshrined at Stanford and later spread across the globe, becoming half of what we know as Silicon Valley values. Only half because McCarthy had no role in that other aspect of Silicon Valley values—the belief that hackers' ideas are best spread through the marketplace. McCarthy was no entrepreneur, and periodically he felt compelled to explain himself. That same fortieth-anniversary celebration featured a panel of fabulously wealthy entrepreneurs who studied computer science at Stanford (and often didn't graduate). McCarthy took exception to the idea that "somehow, the essence of a computer science faculty was starting companies, or at least that that was very important." How could this be true, he asked the audience, since he himself hadn't started any companies? "It's my opinion that there's considerable competition between doing research, doing basic research and running a company," he said, adding grumpily, "I don't expect to convince anybody because things have gone differently from that."[93] To understand why things have gone so differently, we must look elsewhere on the Stanford campus.

2. FREDERICK TERMAN

"Stanford can be a dominating factor in the West"

When McCarthy arrived in 1962, Stanford was fully in the thrall of Frederick Terman, an engineering professor who had been given broad powers as provost to take the university to new heights in higher education. To MIT heights. To Harvard heights. Dragged by the scruff of the neck, if need be. The stakes had always been quite clear to Terman. Two decades earlier, in a letter to a prominent Stanford fund-raiser, he wrote with an engineer's precision: "We will either . . . create a foundation for a position in the West somewhat analogous to that of Harvard in the East, or we will drop to a level somewhat similar to that of Dartmouth, a well-thought-of institution having about 2 percent as much influence on national life as Harvard. Stanford can be a dominating factor in the West, but it will take years of planning to achieve this."[1]

Terman's strategy for Stanford's ascent can be summed up by a single word: entrepreneurism. He wanted the university's faculty and students to think creatively about how to bring in money to the university. When Stanford was flush with cash, there would be scholarships to attract the best students, higher salaries to poach star faculty members from other schools, investments in top-flight facilities befitting such talented students and faculty members. With enough money, Terman concluded, Stanford could overcome its less-than-elite reputation and become the Harvard of the West. Naturally, the number-crunching engineer had a plan to speed things along. Terman identified the various sources for Stanford to tap—primarily government scientific agencies and private businesses—and then made sure that the university pointed its students and faculty members in the right direction with their research.

The way Terman saw it, the benefits to Stanford could flow in any number of ways. There were the significant payments from the government to cover overhead costs at the research labs; the expensive equipment donated by companies eager to encourage research related to their specific technologies; the experts from industry who, for similar reasons, became visiting professors in Stanford's science departments, their salaries paid for in part by their employers; and, more broadly, there was the goodwill Stanford earned from alumni and faculty members who had achieved business success and would give back to the school with hefty donations. The happy coalition of academia, government, and private industry that Terman proposed has been called "the military-industrial-academic complex"; he preferred to call it "win-win-win."[2]

No doubt, the government held the largest pot of money for a university eager to grow. World War II had demonstrated over and over the importance of scientific research: the early computers helped crack the German codes; radar bolstered the Allies' air defenses, while radar-jamming devices thwarted the Germans'; the proximity fuse made bombs more accurate and thus more destructive. Above all else was the atomic bomb, which brought the war to a hasty conclusion, saving the lives of hundreds of thousands of American soldiers, many of whom later became members of Congress. Scientific research would evermore be at the heart of America's defenses, organized around university laboratories, and Terman made sure that Stanford was included among the universities destined to receive those grants. During the 1950s, government-sponsored academic research had more than tripled, reaching nearly $1 billion by the end of the decade—more than three-quarters of those funds went to just twenty elite universities, which thanks to Terman now included Stanford. By 1960, 39 percent of Stanford's operating budget came from federal support, the overwhelming majority directed at physics and engineering.[3]

But Terman's innovation was in converting Stanford research into business opportunities for students and faculty members. The marketplace represented a pot of money for Stanford to tap, too, and Terman wasn't about to ignore it. These ideas first occurred to him in the 1930s in his own engineering lab on campus. Terman's electronics research had practical applications for devices like radio-wave amplifiers and bomb fuses, and he eagerly worked with businesses—established ones as well as start-ups—to bring these products to market. Famously, in 1939 he helped stake two promising former students, William Hewlett and David Packard, in a start-up run out of a Palo Alto garage. In part, Terman was responding to the Great Depression, when even his best students were having a hard time finding steady work. Being in partnership with businesses, or starting your own, was an excellent way to ensure employment immediately after graduation. But he was also thinking of Stanford. Hewlett, who along with Packard became a lifelong Stanford donor, recalled a long-ago conversation: "We were walking out of the old engineering building, and Terman said he was looking forward to the day when I gave my first million dollars to this laboratory. I remember this, because at the time I thought it was so incredible."[4]

Later, when Terman became university-wide provost in 1955, he applied this same strategy to build up Stanford's status through efforts like the Stanford Industrial Park, which brought dozens of new and established high-tech companies into the bosom of the university.[5] He consciously, and unapologetically, steered Stanford toward academic disciplines like biochemistry, statistics, and aeronautics, which were likely to generate business opportunities or government grants, and away from dead-end fields like taxonomy or geography or classics. A later Stanford president, Richard W. Lyman, marveled at how Terman could display such cool confidence while "downplaying or even eliminating established programs or academic emphases that lacked promise for the future."[6] Terman did, however, shed his calm demeanor when challenged over his plans for reinventing Stanford's academic priorities. His typical response to those who stood in the way was to question their competence, motives, or intelligence.

For example, there was a prominent Stanford naturalist, George Myers, who criticized Terman for shifting the biology department away from its strength in zoology and the environment toward research done in laboratories, which was more likely to be marketable. Myers went over Terman's head to complain to the university president, J. Wallace Sterling, about his provost's obvious disdain for scientists "who operate in muddy boots." Despite Terman's insinuations, the nature sciences were eminently practical fields, Myers wrote, addressing humanity's most pressing concern: "The destruction and poisoning of man's complex biological and physical environment by man himself."[7] Sterling backed up Terman, as he always did. Terman never directly engaged with Myers but did privately dismiss his complaints as coming from "a hardworking but not particularly bright biologist . . . who specializes in fish."[8]

Seemingly less controversial was Terman's decision to shrink the classics department at Stanford by failing to replace two professors who had retired. Who could dispute that the department failed to pull its weight financially? Not that Terman was blaming the faculty members for their lack of entrepreneurial zeal: business opportunities from the study of Latin and Greek were painfully small. Still, facts are facts. Why would Stanford direct its limited resources toward a field with such a small potential payoff? When the senior member of the department, Hazel Hansen, a specialist in ancient Greek pottery, wrote to object to the unfilled vacancies, Terman was dismissive. He didn't reply, but again disparaged privately: Hansen, a 1920 Stanford graduate who had risen from lecturer to full professor, was a "single woman—lonely—frustrated."[9]

Opponents of Terman's plans never managed to gain any traction. They were isolated, and he had the numbers on his side. He knew more about most departments than the chairmen did, and he proved it with detailed spreadsheets and charts that tracked, say, the number of graduate students supervised per faculty member, or the average number of years a graduate student spent to complete a PhD. Collecting the relevant data became his personal obsession. Faculty members recall graduation ceremonies attended by Terman: there he'd be, sitting quietly on this day of celebration, with a mechanical pencil in hand, "carefully using his program to calculate and compare doctoral production statistics."[10] His policy was to insist that budget requests arrive on Christmas Eve, "so that while others were celebrating he could get a head start on the next year's work."[11]

For our story, Terman's role was vital. He laid the groundwork for a new relationship between a university and business, which proved particularly relevant when computer science was later shown to have the potential to extract fortunes from the American economy. Because of Terman's ideas, improved techniques for indexing the Web (Yahoo), or searching the Web (Google), or sharing photos online (Instagram)—three among thousands of business-ready ideas developed on the Stanford campus—didn't remain there as part of some free, public trust. (Google, in particular, seemed well on its way to a noncommercial future if not for the pull of Stanford's entrepreneurism.) Instead, all of these start-ups were aggressively brought to market, where they have become central to our lives and hugely valuable assets.

Fittingly, Stanford itself was an ambitious start-up backed by a wealthy couple, Leland and Jane Stanford. The Stanfords planned to reinvent higher education by investing the millions of dollars they had made by connecting Americans through the new transcontinental railroad. As Leland Stanford explained: "If I thought that the university at Palo Alto was going to be only like the others in our country, I should think I had made a mistake to establish a new one, and that I had better have given the money to some existing institutions."[12]

The creation of Leland Stanford Jr. University represents the triumph of hope over overwhelming grief. The Stanfords' son, Leland Jr., was nearly sixteen in March 1884 when he died of typhoid fever during a European tour. The story goes that the father was in such despair at the bedside of his dying boy that he questioned whether he should go on living, and his son appeared in a dream to say: "Father, do not say you have nothing to live for. You have a great deal to live for. Live for humanity." When Leland Sr. awoke, the boy had passed away, and the Stanfords had a new purpose in life.[13] They considered building a combined museum and lecture hall in San Francisco, but concluded that their idealistic son who loved learning would have wanted them to create a university to make higher education accessible to the masses, men and women, rich and poor.

Back east, they sought an audience with the formidable president of

Harvard University, Charles William Eliot. How much money, they asked, would it take to create a world-class university? Eliot's response was that they should be prepared to spend at least $5 million, a sum that was significantly more than Harvard's endowment at the time and perhaps was meant to put a scare into his visitors. Eliot recounted what happened next: "A silence followed, and Mrs. Stanford looked grave; but after an appreciable interval, Mr. Stanford said with a smile, 'Well, Jane, we could manage that, couldn't we?' And Mrs. Stanford nodded."[14] Clearly Eliot hadn't a clue about the kinds of fortunes there were out west even then.

Leland Stanford was worth tens of millions of dollars when he met with Eliot. Though the loss of his son changed his perspective radically, Stanford continued robber baron–ing, simultaneously running the Central Pacific Railroad, the western end of the transcontinental railroad, while also holding political office. Decades earlier, Stanford had been California's governor; during the years the university was being built he was a U.S. senator. Much of Stanford's fortune was in the form of vast tracts of land he had acquired in the course of expanding his railroads; they were among his first gifts to the university. An 8,180-acre parcel in the Santa Clara Valley, the so-called Farm where Leland Sr. would graze his horses, and where Leland Jr. would take his lunches, was set aside for the campus and according to the founding grant could never be sold.[15]

The Stanfords' plans for a university "created something of a sensation in the insular world of American higher education" when they became public in 1885, both for the huge sums of money involved and the new approach—there would be equality between the sexes, a nondenominational religious life, and free tuition.[16] Stanford University's pledge in its founding document to provide an education that led to "personal success and direct usefulness in life" was a head-on critique of East Coast schools whose students "acquire a university degree or fashionable educational veneer for the mere ornamentation of idle and purposeless lives."[17] Leland Stanford promised he would spare no expense to remake education in this way. Pocketbook out, he first searched for a suitable university president. "The man I want is hard to find," he said. "I want a man of good business and executive ability as well as a scholar. The scholars are plentiful enough, but the executive ability is scarce."[18]

In what would become a familiar pattern, Leland set his eyes on the president of MIT (then known as Boston Technical Institute), Francis A. Walker, a Civil War hero and economist. During visits east, Stanford wooed Walker and promised to increase his salary several times over, but never could persuade him to relocate from Cambridge to the frontiers of California.[19] The Stanfords instead settled on David Starr Jordan, a young nature scientist who was president of Indiana University. Jordan was educated at Cornell, something of a model for the Stanfords in that it was cofounded in 1865 by a successful businessman, Ezra Cornell, with the purpose of expanding access to higher education. Up to that point, there hadn't been universities supported by a single wealthy patron. The largest donation to an American college was $50,000 given to Harvard by Abbott Lawrence in 1847 to create an engineering school. In rapid succession after Cornell came multimillion-dollar donations to Johns Hopkins, Stanford, the University of Chicago, and Duke, among others.[20]

"Inevitably, the increase in the size of gifts changed the relations of donor to recipient," wrote the historians Richard Hofstadter and Walter P. Metzger. "Borrowing a term from economic history, one may say that the givers became entrepreneurs in the field of higher education. They took the initiative in providing funds and in deciding their general purposes."[21] Even more than other founding donors, the Stanfords made clear that their grant to the university came with strings attached. The authority one would expect to reside in a board of trustees or president would reside in them, including the power to hire and fire faculty members and to make strategic plans. Jordan was expected to go along with this new way of doing business and in return he was promised that the faculty he recruited would have "at least a little more than is paid by any other university in America."[22]

When Stanford opened its doors in 1891, after six years of planning, many feared that no students would show up, even if the tuition was free. Instead, Stanford enrolled 555 men and women, who were taught by a faculty of fifteen that grew to forty-nine by the second year.[23] Among the members of that pioneering class was the future U.S. president, Herbert Hoover. But despite this fast start and fabulously wealthy patron, Stanford endured a hard few years: "That the infant university survived was something of a miracle," one university historian concluded.[24] In 1893, before the first class had graduated, Leland Stanford died, depriving the university of a powerful protector. His estate was tied up in court while the federal government pursued millions of dollars of what it said were unpaid loans. There was an economic panic that year as well, which threatened the economic growth in the west and made the fight over Leland Stanford's estate even more hard-fought.

Jane Stanford stepped up to keep the school afloat during this period of economic uncertainty, ignoring business managers who recommended that she allow the school to shutter until she could regain her financial footing.[25] Instead, she convinced the judge presiding over the claims against her husband's estate to allow her to maintain the usual household expenses—$10,000 a month to operate three large homes with seventeen servants. She then petitioned the court to consider the Stanford faculty as her "servants." By reducing her household staff to three, and expenses to just $350 a month, she was then able to direct $9,650 a month toward university salaries. In letters to President Jordan, Stanford made clear the depths of her suffering: "I have curtailed all expenses in the way of household affairs and personal indulgencies. Have given up all luxuries and confined myself to actual necessities." Her visits to New York City, where the Stanford estate was being litigated, were an ordeal as well: "It is no pleasure to be on the fifth floor of the Fifth Avenue Hotel in a small bedroom opening on a court, economizing almost to meanness, when I have a sweet beautiful home to go to."[26]

In 1898, when Leland's estate was freed from its entanglements, Jane Stanford transferred $11 million in assets to the university, out of what would be a total of $20 million in gifts.[27] After making such deliberate sacrifices on behalf of her school, which began as a surrogate for her lost only child, Jane Stanford expected unquestioning loyalty from Jordan and the administration. Thus the university was transformed, in the words of Hofstadter and Metzger, from "this unusual oligarchy into a still more unusual matriarchate."[28]

Intriguingly, this matriarch's greatest influence on Stanford came through her insistence that the university cap the number of women who could attend. The trend line quickly grew worrisome to Mrs. Stanford: 25 percent women in the student body when the school opened its doors in 1891, 33 percent four years later, when the first class graduated, and 40 percent in 1898. There were other troubling indicators that year too. Stanford was routed in athletics by its rival, Berkeley, and the student yearbook, weekly newspaper, and debating team were all led by women. Mrs. Stanford asked openly the question that weighed on her mind: How could such a feminine institution, the soon-to-be Vassar of the West, be a fitting final tribute to her son?[29]

The next year, Jane Stanford decreed that the university could never have more than five hundred women enrolled. It was a draconian step, taking no account of how Stanford's enrollment might grow, and entirely out of keeping with the progressive founding mission.[30] Back in 1885, at the first meeting of the university's board of trustees, Leland Stanford said an education "equally full and complete" was of the highest importance. In a newspaper interview four years later he went one step beyond equality between the sexes and said, "I am inclined to think that if the education of either is neglected, it had better be that of the man than the woman, because if the mother is well educated she insensibly imports it to the child."[31]

President Jordan was sympathetic to Mrs. Stanford's concerns in this matter, particularly the noticeable decline in Stanford's sporting prowess. "The university must have more men if we expect to make a better showing in athletics," he said, but he asked that Jane Stanford be reasonable and impose a percentage cap, instead of a fixed number. Mrs. Stanford's response was clipped and impatient: "To carry out this idea would be violating my instructions and very disappointing to me."[32] In later years, administrators employed various work-arounds to keep talented women in the fold, such as letting them audit classes while they waited for one of the five hundred slots to become available, but only after more than thirty years did the board of trustees relax the limit on women to 40 percent of total enrollment. Not until the 1970s was any limit on women's enrollment removed from Stanford's rules.

There were other interventions by Mrs. Stanford. For example, because of her disapproval of automobiles, aka "devil wagons," there would be no cars on campus during her lifetime.[33] She pushed Jordan to use the university's funds to complete construction of the campus, rather than add more professors, as he preferred. The most controversial action by Mrs. Stanford in her day was to demand that a radical economics professor and friend of President Jordan, Edward Ross, be fired for extreme anti-immigrant comments he made during the 1900 election. She got her wish, but it came at cost. In 1903, in the wake of the "Ross Affair," Jane Stanford agreed to give up personal control of the university, allowing the board of trustees to run it. It's true that she simultaneously joined the board of trustees and was elected its president, but, still, the devolution of power had begun. Jane Stanford was dead within two years, under mysterious circumstances.

Today, historians are confident in saying that Mrs. Stanford was poisoned, the victim, in fact, of two separate attacks barely a month apart, the first in San Francisco, the second in Hawaii, where she traveled to recover from the first poisoning. But because Jordan acted quickly, immediately traveling to Hawaii when he learned she had succumbed there, he convinced the national press that the cause of death was heart failure. Jordan disparaged the Hawaiian doctor who treated Mrs. Stanford as she was dying, Francis Howard Humphris, as "a man without professional or personal standing," in order to cast doubt on his conclusion that her death was a homicide.[34] However, an investigation of the death a century later by a Stanford medical professor, Robert W.P. Cutler, concluded that Humphris offered excellent medical care: he did all he could to save Mrs. Stanford's life, and his judgment that she was poisoned was spot on too.

Jordan managed to avoid a public relations nightmare, but whose reputation was he protecting, the university's or his own? "The question need not be skirted," writes another Stanford professor who has studied the case, W.B. Carnochan, about whether Jordan killed Jane Stanford to free himself from her meddling. "He had the motive . . . but could he have brought it off without help from a pharmacist and a household servant? Even the most vivid conjecture resists the notion of Jordan slipping into Mrs. Stanford's San Francisco pantry, bath, or bedroom to spike her mineral water or lace her bicarbonate of soda with strychnine."[35] Mrs. Stanford's personal secretary, Bertha Berner, a beloved long-term employee, "seems to have had ample opportunity but no obvious motive," Cutler concludes. "Jordan seems to have had motive but no obvious opportunity."[36] There simply isn't enough evidence, both scholars say, to reach a conclusion.

———

After Mrs. Stanford's death, Jordan remained president of Stanford for eight years, free of her second-guessing. He expanded the faculty by recruiting scholars in his mold—young ambitious midwesterners willing to help build up a school that was new on the scene. Among that wave was Lewis Terman, a junior professor of educational psychology born to a large family on a rural Indiana farm and educated at Indiana University. Terman arrived in 1910 to a campus that was still repairing the damage from the earthquake four years earlier; he was joined by his wife, Anna; ten-year-old son, Frederick; and seven-year-old daughter, Helen. As a Stanford professor, Terman planned on continuing his studies of human intelligence: What is it? How do we measure it?

Terman's friend and colleague, Edwin G. Boring, offered a facetious definition: intelligence was the ability to do well on intelligence tests.[37] Obviously, Terman believed a well-constructed test was more significant than that. Building on the work of the French education reformer Alfred Binet, Terman devised the first popular intelligence test in America, what came to be known as the Stanford-Binet IQ Test, a standard tool in education, government, and industry.[38] Once he had IQ scores that he believed reflected innate intelligence, Terman could then test his theories. He could, for example, compare fathers to their sons, to study whether intelligence was inherited, or compare members of one race to those of another to determine if one was superior. He could correlate intelligence to other measurable qualities to explore a theory like the smarter you are the more money you earn, or the better grades you get, or the fewer years in prison you serve.

In his long Stanford career, Terman researched every one of these questions, concluding, among other things, that intelligence indeed was inherited and that the white race had more of it than the black race. But it was that last question—what benefits accrue from being intelligent?—that most preoccupied Terman. He wanted his research to reverse the public's perception of geniuses (male geniuses, actually) as peculiar, even deviant, weak, feminine, socially awkward. He wanted to prove what he knew firsthand, that brilliant children grow up to be highly productive, brilliant adults, if only they are recognized and given opportunities. Clearly, Terman's own life was a motivation.

Despite his evident talents, Lewis Terman had seemed destined for a life on the farm. Through exceptional hard work, however, and a powerful drive to escape rural life, Terman arrived at a bare-bones teacher's school before transferring to Indiana University to complete his undergraduate education and then a master's degree in education. He later earned a doctorate at Clark University in Massachusetts, which had built a reputation as a leader in the relatively new field of psychology. Terman's feeling of intellectual inadequacy never quite left, and he devoted his professional life to finding all the little Lewis Termans out there and encouraging society to support them while they were still young. Smart kids shouldn't have it as hard as he did.[39]

At Clark, Terman first experimented with ways of testing for intelligence, recruiting seven "stupid" children and seven "bright" children as his subjects.[40] When he arrived at Stanford, he learned of Binet's intelligence test and realized that he didn't need to start at the beginning. He paid a token fee of a dollar to the uninterested Binet for the rights to publish a revision, Stanford-Binet.[41] The purpose of Binet's test, which was created at the behest of the French government, was to identify children who were falling behind so they could receive specialized instruction. Terman in his research projects pointed the test in the other direction, using it to identify brilliant young people, "It is to the highest 25 percent of our population and more especially the top 5 percent that we must look for the production of leaders who will advance science, art, government, education and social welfare generally."[42] Terman also saw the business potential in IQ tests, which were sold to schools, governments, and businesses. According to estimates based on sales of the test record forms, about 150,000 persons a year took the Stanford-Binet test from 1916 to 1937, which increased to about 500,000 persons a year from 1937 to 1960.[43]

Though Binet never spoke up during his lifetime, he apparently disapproved of Terman's amended test and how it was applied. According to Binet's close collaborator in France, Theodore Simon, Binet considered attempts like Terman's to render a verdict on someone's potential based on a test score to be a betrayal of its purpose. More important than intelligence was judgment, Binet and Simon had written a few years earlier, a quality otherwise known as "good sense, practical sense, initiative, the faculty of adapting one's self to circumstances." They concluded: "A person may be a moron or an imbecile if he is lacking in judgment; but with good judgment he can never be either. Indeed the rest of the intellectual faculties seem of little importance in comparison with judgment."[44]

Terman defended his obsession with intelligence in an essay that appeared in 1922 in a popular magazine of the time, *World's Work*. "When our intelligence scales have become more accurate and the laws governing IQ changes have been more definitively established," he wrote, "it will then be possible to say that there is nothing about an individual as important as his IQ, except possibly his morals."[45] The year before, Terman began work on his ambitious research project, "Genetic Studies of Genius." His assistants ultimately found more than 1,500 young people in California who scored at least 135 on Terman's IQ test, where a score of 100 is defined as normal, making them the top 1 percent of the population.[46] He matched the students, who typically were born in 1910, with a similar number of ordinary young people as a control group; Terman's team would check in periodically—the first time was five years later—to see how the geniuses were faring as compared to young people of average intelligence.[47] This research, he suspected, would demonstrate the general superiority of the smartest, not just intellectually, but physically, morally, socially. Terman wanted to reassure the public that the geniuses his test identified were just like them, only more so. And thus most fit to lead.[48]

The years Terman was starting his research were the high-water mark in the United States for eugenics, the movement that applied Darwin's ideas about inherited traits and "survival of the fittest" to humanity. The elites who promoted eugenics, including President Jordan of Stanford, were reacting to the influx of poor European and Asian immigrants and an inarticulate sense that, as a result, the population was becoming less intelligent and less moral. In crude terms, eugenics was a plan to ensure that the "right" people were reproducing and that the "wrong" people were not. Most eugenicists focused on the "wrong" side of the equation, so-called negative eugenics, and used the IQ test to identify the "feeble-minded" who needed to be prevented from having children for the sake of the human gene pool, through increased use of birth control, and, in the most extreme cases, forced sterilization, a practice that was endorsed by the Supreme Court in 1927.[49]

Terman's research emerged from his belief in "positive" eugenics: if he could prove the overall superiority of brilliant men, they would find it easier to attract the best women and hand down superior genetic material to the next generation. Almost from the start, however, Terman's methods were controversial. The critic Walter Lippmann, writing in the *New Republic* in the early 1920s, took aim at a key vulnerability in Terman's project, his IQ test. "It is not possible, I think, to imagine a more contemptible proceeding than to confront a child with a set of puzzles, and after an hour's monkeying with them, proclaim to the child, or to his parents that here is a C- individual," Lippmann wrote. "It would not only be a contemptible thing to do. It would be a crazy thing to do, because there is nothing in these tests to warrant a judgment of this kind."[50]

In a familiar pattern for the Terman family, Lewis responded harshly and dismissively when crossed. His reply, also in the *New Republic*, ridiculed Lippmann's lack of scientific rigor in suggesting that high-quality childcare was the best way to produce smarter and healthier children. If how we raise children really made them smarter, Terman replied, then Lippmann had to immediately share what he knew, for the sake of the nation. "If there is any possibility of identifying, weighing and bringing under control these IQ stimulants and depressors we can well afford to throw up every other kind of scientific research until the job is accomplished," he wrote less than a month later, imagining how "the rest of the mysteries of the universe would fall easy prey before our made-to-order IQ of 180 to 200." In a sense, Terman was mocking Lippmann for claiming he could produce an "artificial intelligence."

Terman couldn't resist continuing his mockery: think of the entrepreneurial opportunities from a Lippmann-licensed childcare system for improving children's IQs. "If he could guarantee to raise certified 100s to certified 140s, or even certified 80s to certified 100s, nothing but premature death or the discovery and publication of his secret would keep him out of the Rockefeller-Ford class if he cared to achieve it," Terman writes, adding cruelly that he knows a rich father who would gladly pay Lippmann "10 or 20 million if he could only raise one particular little girl from about 60 or 70 to a paltry 100 or so."[51]

Terman was much more kindly disposed toward his young

geniuses—"my gifted children," he called them, a phrase he coined. He tracked the young people with high IQs in his study through adolescence into adulthood and beyond, producing four separate volumes during his lifetime. Upon being enrolled in the genius study, the children, who were kept anonymous, provided a raft of information about themselves and their upbringing—their health, family background, family income, home life. This was the data Terman used to demonstrate how well his geniuses fared in life. Indeed, he found that two-thirds of the "Term-ites" earned bachelor's degrees, ten times the national rate, as well as an unusually high number of doctorates and medical and law degrees. The geniuses' earning power was higher too: while the median salary for white-collar jobs in 1954 was $5,800, the equivalent in Terman's group was $10,556.[52]

The study's usefulness, it should be said, was quite limited. There were the flaws with the test itself, and how it was applied: the geniuses discovered by the test were a bit more likely to be male than female, but that was less jarring than the near total absence of blacks, Japanese Americans, or American Indians. Also, incredibly, Terman's team individually tested two California boys who grew up to win Nobel Prizes in physics and neither scored high enough to be included in the study. One of those boys who was particularly stung by being passed over was young Billy Shockley of Palo Alto, whose Nobel Prize recognized his role in producing the first transistor; alongside Lewis Terman's son, Frederick, Shockley would become a central figure in the growth of Silicon Valley.

Beyond the test's dubious accuracy, there was a flaw in how the study itself was conducted over the years: Terman repeatedly intervened in the lives of his high-IQ subjects, often without their knowledge. He couldn't help himself. He wrote them recommendations for admission to Stanford, gave small sums of money during tough times, and, in one case, helped a fourteen-year-old be placed in a good foster home rather than returned to his abusive father. (To be clear, Terman didn't intervene to help any in the control group.) These intrusions by Terman made the study's conclusions scientifically unreliable, to say the least: readers could rightly question if the geniuses outshone the control group because of their talents or because they had an influential guardian angel. For all its statistical charts and plots of standards of deviation, his work wasn't science—it was advocacy. His dream was a society led by a hereditary class of super-intelligent beings. A beehive run by a few king bees.

In summing up Lewis Terman's life, his friend Edwin Boring tried to explain his commitment to the talented 1 percent. "Some persons . . . wonder how such an undemocratic view could be held by this tender-minded, sensitive, ambitious person, but the fact is that Terman thought of the intellectually elite as those who would save civilization for democracy," Boring wrote in a biography of Terman for the National Academy of Sciences in 1959. "The gifted were given. You do not choose to have them, for there they are, whether you will or no. You can, however, choose to use them, to separate them from the crowd so that they may be trained to devote their special talents to benefit the crowd from which they have been taken."[53] Lewis Terman, patron saint of the Know-It-Alls.

Frederick Terman grew up on the Stanford campus idolizing his father, who in time became the chairman of the psychology department. The feelings between Lewis and Fred were no doubt mutual—the son had pleased his father with his overall academic success. A classmate's caricature, which appeared in the 1916 Palo Alto High School yearbook, showed Fred in academic dress with a book planted right in front of his face. He had other interests as well. He ran track in high school and at Stanford, and was obsessed with ham radio—the personal computer of its day. When Fred graduated from Stanford, first with a bachelor's degree in chemistry, then an advanced engineer's degree in 1922, he had nearly perfect grades. Despite the family's deep Stanford loyalty, both father and son agreed that Fred should head east for the sake of his career. He was accepted at MIT, where he earned a PhD in electrical engineering under the tutelage of influential professor Vannevar Bush. After completing his PhD, Terman was offered a teaching position at MIT and was tempted to stay. A bout of tuberculosis, best treated in the climate back home, was enough to end that flirtation.[54]

With a year of recuperation behind him, Fred Terman accepted a full-time appointment at Stanford, where he taught the science of radio broadcasting. Terman negotiated for the big radio companies to donate equipment to his lab so students could carry out practical scientific work. "Even Terman's textbook, *Radio Engineering*, reflected this strong commercial bent," one historian of the Cold War university wrote, noting, "It became an immediate bestseller because, like his courses, it placed real-world problems at the center with an elegance and simplicity that especially appealed to working engineers."[55] Around this time, Terman began courting Sibyl Walcutt, who was a graduate student in Lewis Terman's psychology lab, where she naturally would have been expected to take an IQ test. Son was no different than father when it came to valuing intelligence. "The young professor went over to the Psychology Department and looked up her IQ scores," Terman's biographer writes, "then, according to Sibyl, things heated up."[56]

During World War II, Fred Terman again traveled east, this time with his family, wife, son, and daughter. He was sent to Harvard by Professor Bush, who supervised the government's wartime research as the chairman of the Office of Scientific Research and Development. Terman's mission there, as head of the Radio Research Laboratory, was to find countermeasures to Axis radar systems. By 1943, the lab had a staff of more than eight hundred and a budget bigger than all of Stanford University. Terman made sure to include Stanford scientists in his plans—more than thirty graduate students and faculty members crossed the country to work at the lab, learning firsthand the kind of research being conducted back east.[57]

When Terman returned to Stanford for good, as new dean of the engineering school, he was ready to lead, confident in his managerial skills and his ability to bring lucrative military research grants westward.[58] At the same time, Terman continued the business-friendly policies he began before the war. The success of Hewlett-Packard had great personal meaning for Terman, but more importantly it was proof of concept. In the 1950s, Stanford began a series of programs to formalize the strong ties between businesses and the university. One idea was to recruit visiting professors from the labs of private companies, so that Stanford faculty and students could learn about the latest trends in industrial research, while the university paid at most half the salary. Another, the Industrial Affiliates Program, made sure that the university's ideas flowed in the other direction by granting prepublication access to Stanford research in exchange for substantial fees.

Each policy gave Stanford the resources to keep growing, as Terman insisted, while creating the mutual dependence between the university and nearby companies that he craved.[59]

The construction of an industrial park on university land in the mid-1950s gave these relationships a sense of permanence. Stanford labs and industrial labs would be cheek-by-jowl. Many of the firms at the industrial park had explicit Stanford ties—Varian Associates, Hewlett-Packard, Shockley Semiconductors. Others were branches of industry leaders like Lockheed and General Electric. Truth be told, the industrial park wasn't devised by Terman, or run by him. Other administrators had promoted an industrial park as a way of extracting revenues from the prized land, the Farm, which couldn't be sold. The combination of the industrial park and a shopping center with half a million square feet of retail space, both of which operated under ninety-nine-year leases, brought in nearly $1 million a year in revenue to the university.[60]

Stanford's tight embrace of individual businesses, even as it was subsisting on government research grants, perhaps should have raised more concerns. Tax money was enriching investors, but these were different times. "Cold war rhetoric," writes the historian Rebecca Lowen, "linked economic prosperity and military might as the two pillars of America's defense against the Soviet threat. Framed in this way, with the emphasis on the 'national good,' the fact that private companies were profiting from the expenditure of public funds could go unremarked."[61] Thriving industry meant thriving country. And thriving industry required well-trained experts, which is how universities made their contribution to industry, with the help of the federal government. When Fred Terman praised this system for bringing technical skill to big business in a 1956 address to engineers there were clear echoes of Lewis Terman's ideas about ensuring that the smartest are in charge. "The idealists, the social planners, the do-gooders, the socialists and others of their ilk . . . called for better distribution of wealth," Frederick Terman said, while engineers, working within the system of free enterprise, simply got it done, "making possible the creation of so much new wealth that redistribution was unnecessary."[62]

In 1955, Terman was promoted to university provost by President Sterling and three years later added the title of vice president of academic affairs after the consulting firm McKinsey and Company recommended a reorganization of Stanford's administration that gave Terman "responsibility for the affairs of each of the university's schools, libraries and eventually all institutes."[63] The brains of the operation. The engineering school was making gains under Terman, but the university as a whole was in decline. The war years had been harder for Stanford than other universities because of the vestiges of Mrs. Stanford's quota on women. In 1933, the board of trustees had relaxed Jane Stanford's strict limit of five hundred women to a 40 percent cap on women, but when war came Stanford was stifled from admitting enough women to fill its classrooms. As a result, Stanford's salaries were not keeping up with the competition back east—in 1954, 75 percent of full professors there made less than the minimum salary at Princeton.[64]

Terman had a mandate to apply his engineering school ideas more broadly. Not only would Terman pour resources into lucrative research areas, but he would recruit only the top men in those fields. He called this strategy "steeples of excellence." The same efficiency argument for academic departments—ecology bad, biochemistry good—would apply to individuals. What made this strategy even more compelling was the decision by Vannevar Bush, as the government's civilian science chief, to automatically include within research grants a percentage to cover "indirect" costs, the university's overhead.[65] Professors who could win large research grants weren't just helping themselves, they were helping Stanford compete with other big-name schools. As to complaints that Stanford was abandoning huge swaths of knowledge under this scheme, Terman replied that such a responsibility fell on schools like Harvard and Yale, which could afford it.[66]

The main challenge for Terman under "steeples of excellence" was how to locate and lure these star professors. First, he evaluated the talent. No matter what the field was, Terman felt empowered to ask, "Is this person smart enough to be a Stanford professor?" In several cases, he discovered that candidates already approved by departmental hiring committees had received poor grades in calculus while in college and declared them unfit. When a candidate had been agreed upon, Terman would look for a way in—whether through a steep raise or speeded-up tenure or by exploiting the internal politics at the candidate's school. Planning trips to Stanford in the winter for candidates from the Northeast was another tactic.

The hiring of John McCarthy is a case in point. As we've seen, McCarthy had reasons to leave MIT, but he was still raw over the earlier rejection by Stanford as a junior mathematics professor, so much so that when the university's only computer science professor, William Forsythe, first called McCarthy about joining him, McCarthy was dismissive. "I thought to turn him off by saying, 'I'd been to Stanford before. I'd only come as a full professor,' and he said, 'I think I can arrange that,' which surprised me very much, since MIT had just made me an associate professor."[67] Computer science had become important to Terman, as he told his top aide: "We got to do this. I see a field coming up and Stanford's got to be in that."[68]

Indeed, in the decades that followed, microprocessing chip makers like Fairchild and later Intel flourished in Silicon Valley, right on Stanford's doorstep. A thriving community of venture capitalists grew there too, along Sand Hill Road. Thanks to Terman, Stanford would be poised to take advantage of the computer revolution. His market-obsessed thinking is now standard at Stanford: the outgoing president, John Hennessy, was a professor of electrical engineering and computer science who spent a sabbatical year launching a successful chip-manufacturing start-up. Hennessy's successor, Marc Tessier-Lavigne, is a neuroscientist who had been an executive at the biotechnology firm Genentech.[69] Since 1970, Stanford has licensed 3,500 inventions, across a range of fields, generating about $1.5 billion in revenue.[70] A large chunk of that total came in 2004, when the university pocketed $336 million for the stake in Google it was given in return for a license of the underlying search technology Sergey Brin and Larry Page developed there.[71]

This wealth has flowed back to Stanford in other ways, as Terman predicted, including a crown jewel engineering center built in 1977, which was named in Terman's honor and funded by Hewlett and Packard at a cost of $9.2 million. In 1982, at age eighty-two, Terman died in campus housing after witnessing this fitting campus tribute, embodying all he preached about Stanford's path to success. But what came nearly thirty years later, in 2010, was perhaps the truest vindication of Terman's ideas. In that year, Stanford built a new science and engineering quad, which included a new engineering center, named in honor of a different Stanford figure, Jen-Hsun Huang, who has a master's degree in electrical engineering and cofounded the tech company Nvidia. Huang and his wife donated $30 million to the project.[72]

The next year, the Terman engineering center was demolished, with as much as 99 percent of the building material recycled: Spanish clay roof tiles were carefully removed and incorporated in a new outdoor education and recreation center; the cedar planks in the roof were sold to Stanford students and faculty members for use in school projects.[73] Terman is not entirely forgotten: inside the Huang center, on the second floor, is where you'll find the Frederick Emmons Terman Engineering Library.[74]

3. BILL GATES

"Most of you steal your software"

I f the Know-It-Alls' values represent a merger of a hacker's radical individualism and an entrepreneur's greed, then only one man can be considered their forefather: Bill Gates. Before he arrived on the scene, the hackers were largely unchallenged in their instinct for sharing what they had learned and general disdain for making money from the computers they loved. Certainly, their pied piper, John McCarthy, shared this inclination. Gates, however, was prepared to call his fellow hackers out—call them thieves, even—as he built the first great software empire and became the prototypical hacker-entrepreneur.

The story begins one winter day more than forty years ago, as Gates and his high school friend, Paul Allen, were walking in Harvard Square. They spotted at a newsstand the cover of the January 1975 issue of *Popular Electronics*, which featured the Altair, the first build-it-yourself computer, under the banner headline, "Project Breakthrough!" The news was simultaneously inspiring and troubling to Gates and Allen.[1] The arrival of Altair's relatively inexpensive "kit computer" vindicated their shared conviction that a computing revolution was soon to come. Intel had just introduced a microchip so dense with circuits that it could contain the brains of an entire computer, a breakthrough with the potential to allow computers to piggyback on the advances in microchips and become smaller, faster, cheaper. Soon enough, personal, too. All very exciting. The worrisome part came from the recognition that others had the same intuition about personal computers and, what's more, some had already acted. What if Gates and Allen were too late to the party?[2]

Gates was unusual among his hacker peers: he cared deeply about being a business success. His interest in making money could be traced to early childhood, when he pored over issues of *Fortune* magazine, and became evident after he transferred to Lakeside, a private boys' school outside Seattle, as a junior high school student. At Lakeside, Gates was introduced to Allen, who was two years older; they both excelled at computers and learned to program in the BASIC language. In 1968, the school could communicate via teletype with a remote computer owned by General Electric, using a version of McCarthy's time-sharing system. Access fees approached $10 an hour, however, and paying for computer time became a problem almost immediately. Gates and the others quickly convinced the Lakeside Mothers' Group to contribute some of the proceeds of an annual rummage sale to their new computer club. Those funds, too, were used up almost immediately.[3]

Feeding their computer habit would be an obsession of Gates and Allen throughout their teenage years. One way of ensuring access to a computer, the two discovered, would be to go into business. They started a company that used computers to manage a corporate payroll and later were hired by Lakeside to produce new class schedules when it merged with a nearby girls' school. They also created a business that analyzed traffic patterns for the government.[4] "We were kind of desperate to get free computer time one way or another," Gates recalled of his coming of age.[5] A new business in town, the Computer Center Corporation, which was created by University of Washington professors, relieved some of that pressure.

The CCC hoped to expand the market for time-sharing by selling programming time to small businesses. The company, which was housed in a converted car dealership in downtown Seattle, rented a Digital PDP-10, a minicomputer beloved by hackers, and recruited three of the top programmers from John McCarthy's artificial intelligence lab at Stanford. Even so, the time-sharing system was buggy and paying customers certainly weren't as understanding as students when the system crashed and all their work was lost. Too many crashes and the CCC couldn't continue as a credible business. That's where Gates and Allen fit in. They were among a crew of young hackers invited to discover flaws in the time-sharing system. "Having a few of the students, including me, bang on it and try to find bugs seemed like a good idea," Gates explained. "And particularly, let us do that mostly at night. . . . So, for a few years that is where I spent my time. I'd skip out on athletics and go down to this computer center."[6] The students got free computer time, and CCC got an energetic debugging team.[7]

A couple of years later, Gates was a Harvard undergraduate majoring in applied mathematics, playing in a regular dorm poker game, when the Altair news lit a fire under him. Allen, who had dropped out of college and moved to Boston as a programmer for Honeywell, stoked Gates with the cry, "It's going to be too late. We'll miss it."[8] The Altair was already prospering among a core audience of "hobbyists" who were so committed to the idea of owning a personal computer that they didn't mind that they had to assemble the machine themselves.[9] Incredibly, the hobbyists also apparently didn't mind that, at first, "there was really nothing you could do with it," Gates recalled. "There was no teletype hook-up in the early days, there was no software for it. All you could do was use these switches, key things in into this front panel and maybe do a little program that does things in the lights. . . . People just bought it thinking that it would be neat to build a computer."[10]

Gates and Allen recognized that software would be the way to make these computers both more user-friendly and more engaging. This would be the sea change: up until that point computers were expensive, and mainly used by programmers themselves. Sometimes hackers would write useful software and share it with the computer manufacturer, who in turn would share it with other customers. Mostly, though, the manufacturer would just include whatever software it had produced for its expensive machine and a talented programmer would improvise what he needed. With Micro-Soft, Gates and Allen proposed a different model: users would pay for software that was written to a professional standard by paid programmers.

The first program their company produced for the Altair was one they knew well, BASIC. As the name suggested, BASIC (Beginner's All-Purpose Symbolic Instruction Code) was intended to allow a user to perform tasks on a computer without deep programming experience. Its creators, two Dartmouth mathematicians, John Kemeny and Thomas Kurtz, devised BASIC in 1964 while working under a National Science Foundation grant to make computing more accessible to undergraduates.[11] Like most software of that era, BASIC was freely distributed. The challenge for Gates and Allen was to create their own version of BASIC that fit the constraints of the new Intel microchip. As hard as the programming might be, the greater hurdle would be gaining access to a computer. This time Harvard University would be the main supplier.

While programming BASIC, Gates camped out in the university computer lab. He shared his access with Allen, who lived nearby. The two, with some hired help, quickly produced a working version of BASIC, which was such a hit that every computer owner had to get their hands on it.[12] Gates recalled, "Before we even shipped BASIC, somebody stole the demo copy out of the van and started copying it around and sending it to different computer clubs. There was a real phenomenon taking place there, right around this Altair computer."[13] Seeing the enthusiasm for BASIC, Gates eventually took a leave to focus on Micro-Soft and never returned. He ultimately obtained a Harvard diploma, however, when the university awarded him an honorary law degree in 2007.[14]

The timing of Gates's leave could appear suspect since it came on the heels of a disciplinary hearing with college administrators. An auditor from the Defense Department discovered how much time Gates had spent programming on the PDP-10 computer, which was provided by the U.S. government, and began to ask questions. Gates's explanations were flimsy: first, he was doing work for his own company, and, second, some of that outside work was carried out by Allen, who wasn't a student. The records of the Gates disciplinary hearing are sealed, but accounts say that the punishment fell short of his being asked to leave.

Undoubtedly, however, the time was ripe for Gates to see how far software could take him. An early problem, as we've seen, was how to manage the enthusiasm for Micro-Soft BASIC. If the plan was to get paid to write the software that all computer owners used, the hackers' insistence on sharing software was an obvious impediment. Events were moving fast: only a year after first seeing the Altair on a magazine cover, Gates already had a business that he believed was under attack. He directed his fire at the most influential group of Altair hobbyists of the time, the Homebrew Computer Club, which met every two weeks on the Stanford campus to trade stories about their personal computers and, at times, trade software.

The existence of such a club showed the potential of the Altair—and its inevitable imitators—to disrupt computing, which up to that point had been limited to people who worked at universities, the government, or corporations, or had managed to connect to a remote computer through an expensive time-sharing system. These "homebrewers," however, were the vanguard of what would become the personal computer movement; early members included Steve Wozniak, who would go on to design the first Apple personal computer. As little as these primitive Altair computers could do at first, their owners still enjoyed possessing them, controlling them. After barely a year, the Homebrew club regularly had three hundred members at its meetings on the Stanford campus; six hundred people subscribed to its newsletter.

Gates, who was nineteen years old at the time, made his case in the club's newsletter, which published his "Open Letter to Hobbyists" in the January–February 1976 issue. He didn't mince words. "As the majority of hobbyists must be aware, most of you steal your software," he wrote. "Hardware must be paid for, but software is something to share. Who cares if the people who worked on it get paid?" He then framed the issue in familiar terms of market incentives. That is, if software producers don't get paid, then vital software won't be written: "Who can afford to do professional work for nothing?" Furthermore, there was the computer time, whose "value" exceeded $40,000, Gates informed the hobbyists. This odd phrasing jumped out to some readers, even before people learned of Gates's reliance on Harvard's computers. Gates wasn't claiming that Micro-Soft had spent $40,000 for the computer time, which it needed to recoup; rather, it had obtained that time somehow.[15]

Young Gates certainly had chutzpah. He was writing to a bunch of excited young computer enthusiasts to object to the "theft" of a version of a program that was, to start, based on software created by academic researchers working on a federal grant and was created using "borrowed" time on a university computer that was also paid for by the federal government. The facts may not have been ideal, but Gates nonetheless succeeded in defining the question in a way that played to his advantage. He was insisting that programmers be paid and that computer owners recognize that professional software would make their machines ever more useful.

In computer history, 1976 marks the end of an innocent time. The Homebrew hobbyists had started as a group of barely thirty hackers, yet in no time they had made a sworn enemy, an innocent-looking teenager who was committed to challenging their way of life.[16] While most hackers viewed computers as a great gift to the world that mustn't be sullied by commerce, Gates had seen them only through the lens of business. A fascination with computers frequently had cost Gates money, while at other, magical times, it had sent money his way. When Stanford and the Silicon Valley venture capitalists later insisted that computer innovation should spread through the market and reward entrepreneurs with enormous fortunes, they had to look no further than Gates and Microsoft.

Microsoft's software empire would prove instructive to Mark Zuckerberg as he went about building his social-networking empire, Facebook. Microsoft offered an early example of the potency of the network effect in computing, teaching that if you become the most popular software, new users will sign up just to fit in. The dominant Microsoft operating systems were a source of fascination too. "I thought, you know, building this ecosystem was really neat, and that kind of inspired me. Right?" Zuckerberg said in an interview. "And the way that they built a platform, I kind of thought, 'Okay. Well, maybe one day, you know, the tools that I'm building can be part of a broader ecosystem as well.'"[17]

Under Gates's leadership, Microsoft reached unprecedented levels of profitability for a company its size. Yet Gates didn't allow himself to become quite as utopian about computers as the Know-It-Alls who followed. Unlike the founders of Google, Facebook, Amazon, Uber, and dozens of other Silicon Valley start-ups, Gates never shouted to the world that his company would make the world a better place, in addition to making huge profits. Only after Gates had amassed a fortune did he begin to think in terms of "saving the world," and it wasn't by recommitting himself to Microsoft's success. Beginning in 2006, he shed day-to-day responsibility of the company to focus on the Bill and Melinda Gates Foundation, whose mission is "to help all people lead healthy, productive lives." He recently wrote on Twitter that his nineteen-year-old self's view of humanity was far too limited: "When I left college, there are some things I wish I had known, e.g., intelligence takes many different forms. It is not one-dimensional. And not as important as I used to think. I also have one big regret: When I left school, I knew little about the world's worst inequities. Took me decades to learn."[18]

4. MARC ANDREESSEN

———

"By the power vested in me by no one in particular"

As Bill Gates and others had shown, there were fortunes to be made by an enterprising hacker willing to sell his computer expertise to the general public. But there still was a wide gulf that cut off the tech leaders of this era from social and political power. On one side of the gulf were the leaders themselves, engineers and coders obsessed with computers; on the other were their customers, members of the public or businesses who relied on computers to carry out important tasks but nonetheless thought that having a deep, personal connection with a machine was a bit strange, perhaps even perverse.[1] Tech businesses like Microsoft, IBM, and Hewlett-Packard were already driving the economy, but until this gap was bridged—until regular folks shared a hacker's delight in having a machine fulfill their every whim—these leaders would lack the influence to reshape society according to their values.

This observation may seem, in part, obvious: until computers were intertwined with people's daily lives, improvements in software or hardware wouldn't matter much. But there was another aspect as well. Before there could be a generation of Know-It-Alls to bring Silicon Valley–style disruption to America and the planet, the public's view of hackers needed to radically change. Instead of being feared or pitied, they had to be respected. Their fixation on teaching a machine how to act human—an outgrowth of John McCarthy's original computer-based artificial intelligence project—had to seem benevolent, not menacing or peculiar. In fact, a comprehensive recalibration was about to begin. Not only would hackers soon be admired for their great wealth and ingenuity, but a different class of people—the Luddites who saw computers as a threat to civic life—would take the hackers' place as the new obsessives who lacked basic social skills.

A key marker in the public's changing attitude toward computers—and the young men who loved them—was the arrival of the World Wide Web in the early 1990s. Each of us would grow to love our computers, too. And why not? A computer with access to the Web was nothing less than a revelation, taking you free and easily from essays to stock quotes to pornography to an old forgotten friend to new music to classical literature to video games and back again. A network of clickable hyperlinks propelled a "surfer" along on the Web, an intuitive mechanism for navigating online, especially when compared with the arcane commands and long strings of numbers hackers had used to access the Internet. By this time, there already were paid-subscription services like Prodigy, CompuServe, and America Online, which offered customers a few online tools like email and chatrooms, but the Web was different, promising more and free. Unlike the people who signed up to a subscription service and contributed only to that service or its affiliates, everyone on the Web was pulling together. The collective potential of the Internet, which had largely been hidden from the public since its creation as Arpanet in 1969, would be made plain through the Web.[2]

The first plans for how the Web might work were sketched out in 1989 by Tim Berners-Lee, a consultant at CERN, the European particle-physics research center located in Switzerland. A circumspect, thirty-five-year-old physicist born in Britain, Berners-Lee fit a very different profile than the one we have grown accustomed to among our brash tech leaders. He stumbled on the idea of the World Wide Web, he says, while building a computer database to keep track of the many scientists and support staff members who shuffled in and out of CERN's laboratories. Over time, Berners-Lee discovered that he was more fascinated by the links between individuals than by the individuals themselves. His project, which came to life the following year, was meant to highlight and strengthen the ties among the CERN staff, like a shared language or similar research specialty. "The philosophy was: what matters is in the connections," he recalled. "It isn't in the letters, it's the way they're strung together into words. It isn't the words, it's the way they're strung together into phrases. It isn't the phrases, it's the way they are strung together into a document."[3]

There was a power in this simple philosophy, for it meant that the Web grew thicker and more interconnected every time someone or some group created a new page or simply added a link to a related page. A decentralized network of users, including many not so adept at computers, were creating what was already an unprecedented digital resource. "The Web made the Net useful because people are really interested in information (not to mention knowledge and wisdom!) and don't really want to have to know about computers and cables," Berners-Lee explained.[4] Regular folk, not programmers, were steering this ambitious computer project along, and its designer was fine with that.

In fact, so much about the birth of the Web came, as soccer fans might say, against the run of play in computer innovation. A template had emerged from Silicon Valley: young programmers and engineers, nurtured in the windowless computer labs of America's great universities, came up with projects that they were encouraged to take to market with the backing of venture capitalists. Instead, this most recent breakthrough had bubbled up from a physics lab in the middle of Europe, under the direction of an anti-hacker of sorts who apparently made no effort to profit personally from his discovery!

These developments led to more than a little consternation back in the States, mixed with a determination to get in the game. After all, the hackers were all about how computers, in the right hands, could change the world. A civilization-bending project like the Web was simply too attractive—too potentially profitable—to be ignored and left to its own devices. Users would demand a better system, and Silicon Valley would give it to them. In short order, the Web was pulled toward America, where Berners-Lee's decentralized, noncommercial vision succumbed to the innovative powers of a new generation of hackers-turned-entrepreneurs. First among this generation was Marc

Andreessen, the influential Silicon Valley venture capitalist who at the time was an ambitious twenty-one-year-old computer science major at the University of Illinois, Urbana-Champaign. He was quick to recognize the business potential of a global network like the Web and nearly as quick to act.

Born in 1971, Marc Andreessen grew up unhappily in New Lisbon, Wisconsin, about eighty miles north of the liberal college town of Madison and a world apart. During his childhood in the 1970s and '80s, New Lisbon had a population of about fourteen hundred and was more than 97 percent white. Marc's father, Lowell, was a sales manager for a company that sold genetically modified corn seeds, and his mother, Pat, worked in customer service at, among other places, Land's End. Marc Andreessen describes a life of relative privation—a shared "party" telephone line at home; relatives who had an outhouse; a winter of "chopping fucking wood" when his father decided to stop paying for gas.[5] (His friend and business partner, Ben Horowitz, is genuinely nonplussed by the thought of Andreessen, a large man who stands six foot five, wielding an axe: "It is still hard for me to really visualize Marc chopping wood. It's like asking Einstein to mine coal. How crazy that must have been.")[6]

Andreessen displayed classic "compulsive programmer" characteristics as a child, to use Joseph Weizenbaum's resonant phrase. To start, there was his preteen fascination with the TRS-80 personal computer, which he bought with money saved from mowing lawns, supplemented by a contribution from his parents. Andreessen also had the requisite anti-authoritarian streak. Like John McCarthy at a similar age decades earlier, Andreessen fought a losing battle to be excused from gym class.[7] He thought his public high school was an embarrassment, and didn't hide that opinion. The school had a blandly religious culture, which opposed science, Andreessen recalled, while one history class was "taught out of our teacher's unpublished 800-page manuscript on the JFK assassination conspiracy."[8] There was a computer lab in high school, but it lacked a modem to connect to the wider world of Internet bulletin boards and university computer centers. Young Marc Andreessen was trapped in rural Wisconsin and, alas, even computers offered little help in making an escape.[9]

Yet escaping was precisely what Andreessen's neighbors remembered was on his mind. "I got the feeling that New Lisbon wouldn't keep him," said Paul Barnes, the manager of the local supermarket where Andreessen worked as a bagger and stocker, who recalled being impressed by his employee's large vocabulary and big ideas.[10] Andreessen can rattle off the deprivations of being raised in New Lisbon: the cold weather; the poor diet; the ignorant, superstitious farmers; the husbands waking up early to go ice fishing to avoid their wives; the barren intellectual landscape.[11] In a profile that appeared in the *New Yorker*, Andreessen complained of driving an hour to the west to La Crosse, where all you could find was a Waldenbooks with nothing but cookbooks and cat calendars. "Screw the independent bookstores," Andreessen said in praising the disruption later brought by Amazon .com, which began its march through e-commerce by selling books. "There weren't any near where I grew up. There were only ones in college towns. The rest of us could go pound sand."[12]

Notwithstanding Andreessen's especially dim view of rural life, he still acts as the protector of his former townsfolk, who, as he once expressed in a post to Twitter, are "well aware that the left, intellectuals, politicians, et al look down on them."[13] The chip that Andreessen carries about his rural upbringing—expressed today from the comfort of the Bay Area—raises questions about what drives his enthusiasm for Internet-based social disruptions. Is it faith in the wonderful new world to come, or anger at the hurdles, real and imagined, that he faced as a super-smart teenager growing up so far from the action? If we as a society are going to accept so much disruption and destruction, the assurances that it will be worth all the suffering should come from a place of compassion, not resentment.

Andreessen's first steps toward leaving New Lisbon and finding that action included winning a Merit Scholarship in high school and enrolling at the University of Illinois, where he planned to study electrical engineering. There was nothing romantic or idealistic about these choices, he insists. They were purely mercenary. While in high school, Andreessen read an issue of *U.S. News & World Report* from 1986, which ranked undergraduate majors based on whose graduates earned the highest starting salary; electrical engineering was at the top of the list. The same issue ranked the University of Illinois among the top three schools in electrical engineering, and in short order his college decision was made as well.[14] As to why someone fleeing rural Wisconsin would choose a school in nearby rural Illinois, Andreessen says he thought he was heading to a city of sorts, and then discovered that in Urbana-Champaign, "they had a cow with a hole in its gut so you could see it digesting its food. It was that kind of school."[15]

Soon after arriving at the university, Andreessen concluded that electrical engineering was too demanding and switched to computer science, still something of an obscure discipline in the late 1980s. "Sometimes I just made things up, but then the field was so new, my professors were making things up, too," he recalled.[16] Andreessen found a purpose in programming and was good at it, to boot. He was chosen for a coveted part-time job on campus at the National Center for Supercomputing Applications (NCSA), one of five such centers created in the 1980s by the National Science Foundation.[17] At the center, Andreessen made just $6.85 an hour, but on the bright side he had a desk and an expensive Indigo computer, which he managed to connect to the cable TV box so he could have CNN playing in the background. Around the lab, Andreessen was known as a generally grumpy figure apt to reject tasks as boring or beneath him . . . until that time when a project worthy of his ambitions appeared.[18]

In 1992, the World Wide Web was up and running but was still lacking a browser that worked on all computers, not just the NeXT computer that Berners-Lee first programmed with. A proper browser, which was highly compatible . . . now that was a worthy project! After all, the Web was the new, new thing on the Internet and a browser was the crucial program for the Web—your transportation, your translator, your window, your pad and pencil, your safety blanket. Later, the Web browser would take on more sinister responsibilities—ankle bracelet, chaperone, corporate listening device. But let's not get ahead of ourselves. In that ancient year, 1992, the NCSA was among a number of computer labs quick to take up the challenge of producing a browser that was easy to install, could work on different operating systems, and would improve on the intuitive navigation of the Web.

Andreessen lobbied hard to land the assignment and in the fall he was paired with an experienced staff programmer, Eric Bina, who took on the difficult coding, freeing up young Andreessen to keep his eye on the big picture. "Marc is a strong driving force for changing the world. He is clearly driven to do that," Bina said. He added, by way of contrast, "I don't feel driven to change anything but my own situation."[19] Bina and Andreessen and their growing team of coders worked out of the dark basement offices of the old Oil Chemistry building, offices that soon filled up with piles of pizza boxes, stray cookie packages, empty soda cans, and Skittles wrappers.[20] In a matter of months, they had created a working version of the browser. Significantly, their browser, which was given the name Mosaic, could embed images directly on the page rather than clumsily requiring images to pop up in a new browser window. On January 23, 1993, Andreessen posted a file containing a first working version of Mosaic, under the words, "By the power vested in me by no one in particular, X-Mosaic is hereby released."[21]

Andreessen expected there to be immediate demand for what was, by all accounts, a vastly superior browser. "We just tried to hurry and get it out there, initially to a limited group of 10 or 12 alpha and beta testers," he said. "Of course, the Internet is a great way to distribute viruses, too; put a virus out and then it propagates."[22] Born on a university campus, Mosaic had additional advantages in quickly finding an audience. To start, the campus itself was filled with young people with computers and unusually fast Internet connections who were eager to try something new. When the public did in fact quickly engage with Mosaic—and the original twelve downloads grew to several hundred thousand by December—the team could depend on the university's infrastructure to keep up with demand.[23] No surprise, then, that the most prominent Web browser (Mosaic/Netscape), portal (Yahoo), search engine (Google), and social network (Facebook) all germinated at universities, whether Illinois, Stanford, or Harvard.

The immediate popularity of Mosaic meant that there would be two very different guardians of the nascent Web: Berners-Lee, the scientist who conceived it, and Andreessen, the Midwest-born college student who helped it to catch on quickly. Both men were obviously transfixed by the Web's potential, but if anything, Andreessen was the one more enthralled: great idea, Tim, now let's get on with it. There had never been anything like the Web before; who knew if it would even be popular? Thus the Mosaic team focused on making the Web experience simple, intuitive, and eye-pleasing, starting with the browser's newfound compatibility with images. Andreessen acted more like the leader of a hungry start-up than a member of a university research team. Berners-Lee noticed that after the first version of Mosaic was released, Andreessen maintained "a near-constant presence on the newsgroups discussing the Web, listening for features people were asking for, what would make browsers easier to use . . . almost as if he were attending to 'customer relations.'"[24]

Berners-Lee, by contrast, made a priority of promoting the values of individual autonomy and collaboration on the Web. He insisted that a browser should be a text editor as well, so that Web surfers would be encouraged to add to the interconnections, not just surf across them. Users would have handy tools to create and publish a page on their own or cooperate with friends to write, edit, and publish together. Think of Wikipedia, the online encyclopedia where thousands of contributors create articles individually, but usually improve them collectively, or a shared Google document, which similarly grows as more people are invited to contribute. Berners-Lee wanted these experiences to be the norm. Not just a part of the Web experience, but central to it. This was the democratic instinct as applied to the Internet, with the general public driving the development of the Web, rather than programmers, with all the inefficiency and lack of professionalism that implies, as well as the unpredictability and personal control.[25]

When Andreessen and Berners-Lee finally met face-to-face in 1993 in Illinois, there already was a "strange tension," Berners-Lee reported.[26] The Mosaic team exuded a confidence that they represented the future of Web development, which rankled Berners-Lee. They described material online as being "on Mosaic," rather than "on the Web," another annoying trait.[27] More significant, the Mosaic developers early on dropped the collaborative, text-heavy tools that Berners-Lee championed as empowering the public, seeing them as inefficient and a distraction from the central mission of creating a compelling, entertaining Web experience. With a better browser and faster Internet connections, the Web could become more like the television that Andreessen had already wired into his computer—passive and commercially friendly. But with a crucial difference: a television signal that could reach the entire world at once!

From a Silicon Valley perspective, the Mosaic team was strategically "pivoting" the Web browser toward Andreessen's commercial friendly vision and away from Berners-Lee's, which wasn't. The timing couldn't have been better. By the early 1990s, the last official barriers to business and commerce on the Internet were torn down through a combination of congressional legislation and new rules from the National Science Foundation, the organization that supported the Internet.[28] The noncommercial status of the Internet was rooted in its history as a government-funded project operating mainly through universities and government agencies, but businesses were persistent in arguing that they belonged online as well. In 1993, the Internet became fully open for business with the passage of the National Information Infrastructure Act, which "clearly took the development of the Internet out of the hands of the government and placed it into the hands of the competitive marketplace."[29] This shift didn't necessarily mean that the young programmers like Andreessen who built Mosaic would benefit from its success. They were merely salaried employees at a lab; the University of Illinois retained the rights to Mosaic.[30]

Years later, those early design choices by the Mosaic programming team still made Berners-Lee cringe. "The Web, which I designed to be a medium of all sorts of information, from the very local to the very global, grew decidedly in the direction of the very global, and as a publication medium but less of a collaboration medium," he said in dismay.[31] The experience was a useful harbinger, however. Going forward, the Web experience would largely be in the hands of hacker-entrepreneurs committed above all else to bringing in the most users to the Web, at first to make sure the project would survive, later, to reach profitability. If gaining a huge global audience was your primary goal, even Berners-Lee had to concede, why would you fight for tools to encourage collaborative editing, which "didn't seem to promise that millionfold multiplier"?[32]

Andreessen certainly pled guilty to wanting to please the largest possible audience. "I'm a Midwestern tinkerer type," Andreessen says. "If people want images, they get images. Bring it on."[33] In response to Berners-Lee's other concern, that Andreessen was hijacking control of the Web, the young hacker would turn the question back on him. Andreessen's goal was to share the Web with the world and give users a chance to shape its development by carefully watching which features were popular and which were not, and revising accordingly. He would later press for changes to the browser that helped businesses operate online and, in the process, usher even more users to the Web. Berners-Lee, by insisting that the Web be collaborative and less flashy whether the public wanted these features or not, was the better example of a programmer trying to impose his will on the public. "The Web had already become a brush fire, and he was uncomfortable that he was no longer controlling it," Andreessen said about Berners-Lee in those early days.[34]

Brush fire indeed. Aided by the steady adoption of the Mosaic browser, the amount of information being conveyed by the Web grew more than two thousand times from January 1993 to January 1994, a figure that caught the attention of people attuned to how the economy might be changing, including a young investment banker, Jeff Bezos, considering whether to leave finance to start his own business. "Things just don't grow that fast," he observed.[35]

In December 1993, Andreessen left the University of Illinois to head to Silicon Valley to start earning the high salary promised in that issue of *U.S. News & World Report*. He had just graduated, so this was a natural time to be departing, but he would also be leaving behind Mosaic, the project that had defined his time there. Andreessen's frustration with the Mosaic team had been growing, as the success of the browser outside the lab had caused administrators to take notice.[36] They began to schedule regular meetings to review progress and kept adding members to the programming team. Meanwhile, the university was weighing proposals to license the Mosaic code. The terms of participation were now very different for Andreessen. He had been running the equivalent of a lean, fast-moving Web start-up. Great fun. Going forward, he would be navigating academic turf wars as a recent graduate. Not so much fun. "There was no reason to stay there," Andreessen explained. "The environment was falling to pieces compared to what it had been, simply because there was this influx of money. The project grew from 2 people to 20. It was completely different."[37]

Just before Andreessen left the lab, however, he was given a bloody flag to wave to unite his fellow hackers at the lab. That December, John Markoff wrote a prescient article in the *New York Times* about Mosaic, calling it "a map to the buried treasures of the Information

Age." The big photo accompanying the article featured Larry Smarr, the director of the NCSA, and Larry Hardin, who directly supervised the Mosaic project, but no one else. Neither Andreessen nor Bina, nor any of the other programmers, was mentioned by name.[38] This was the traditional academic model in a nutshell: the students do the work, the professors get their names first on the journal article and in the news media. This public diss would prove quite helpful in the months that followed, as Andreessen tried to lure members of his old team to a new commercial project.

When Andreessen first headed to Silicon Valley in the winter of 1994, however, he had no intention of resuming work on a browser. He joined a small company, Enterprise Integration Technologies, and lasted three months before being recruited by Jim Clark, a former Stanford electrical engineering professor turned entrepreneur. They were to work together on the hot business idea of the time, interactive television. Only when the plan for interactive TV fell through, and Andreessen and Clark were brainstorming ideas, did Andreessen bring up the idea of creating a commercial browser to compete with Mosaic.[39] Clark and Andreessen began hiring as many of the disgruntled NCSA programmers to their new company as possible; they would be joined by a few other early Web programmers as well as some more seasoned hands Clark knew from his previous company, Silicon Graphics. Smarr, who considered Clark a friend, at the time felt betrayed by the "raiding" of the talent at his lab.[40]

Clark, an engineer himself, believed that hiring the best technical talent would be the key to the success of his and Andreessen's new company, Mosaic Communications, MCom for short. Clark flew out to Illinois to close the deals personally. Bina, for example, only signed on after Clark personally agreed to let him work from Illinois so he could stay with his wife, who was a professor.[41] Another important hire, Lou Montulli, a recent graduate from the University of Kansas who had created Lynx, a text-based Web browser, recalled being summoned to Champaign and flying with a last-minute ticket that was so expensive that he wanted to be assured he would be reimbursed for its cost.[42] Nothing to worry about, he was told. Clark's new company would make sure the programmers were appropriately compensated, starting with Andreessen, but extending to the entire programming team. Clark had learned a painful lesson from Silicon Graphics, which grew out of his research at Stanford and was founded with a team of departing Stanford graduate students: over time, the financiers had profited much more from the company's initial success than the engineers, including Clark, even, who at the start sold a 40 percent stake in the company to an investor for $800,000.[43] Ultimately, the investors took over the board and made business decisions that forced Clark to leave.

The MCom team set to work on creating a new browser from scratch, Mosaic Netscape. Clark had raised the idea of paying the University of Illinois a fee to license the Mosaic code, as other companies were doing, but Andreessen said no. Alma mater wouldn't see a penny of MCom money if he had anything to do with it.[44] The programming team would construct a better browser, which would be designed for the slow 14.4 kb modems of the real world, not the fast cables of well-financed universities. What came next was a programming binge straight out of the hacker annals—Montulli, who had extensive responsibilities for coding the browser, painted a picture of programming life at MCom: "Essentially 10 Mountain Dews (full strength, no diet), horrible food and I think my regular schedule back then was to come in, work for 20 hours straight, we had a futon room, which is a little disturbing to think about now, it was a mattress in a conference room that was dark, I would catch 4 or 5 hours of sleep at the office, wake up, do another 20 hours, and then go home and sleep for 12 or 15 hours and start the whole cycle again."[45]

Everything about the MCom work environment resembled the hacking days at MIT, not just the working hours and the bathing habits, but also the near-complete absence of women in any meaningful role. A 1994 Web page presents the team at the time, a total of twenty-three members, some with short descriptions like "Marc Andreessen—the Hayseed with the Know-How" or "Jim Clark—Uncle Jim's Money Store." Seemingly, not a woman among them. The motto of the group, taken from Sartre's *Being and Nothingness*, is filled with collegiate angst: "All human actions are equivalent . . . and . . . all are on principle doomed to failure."[46] The Web, however, seemed to defy gravity or entropy, growing frantically—roughly 600 Web sites at the beginning of 1994, became 10,000 at the end of the year, became about 100,000 by the end of 1995.[47] By late 1994, the original Mosaic browser had an estimated 3 million users. In the next year, 1995, about 18 million American homes had a computer with a modem for connecting to the Internet, an increase of more than 60 percent from the year before, and, for the first time, a majority of Americans used a computer either at home, at work, or at school.[48]

In the fall of 1994, Andreessen was invited to explain the burgeoning online ecosystem to a San Francisco conference for entrepreneurs eager to learn about Web commerce. He was twenty-three and still relatively new to Silicon Valley. His company's built-from-scratch Mosaic Netscape browser had just been released. Working with an overhead projector and a bunch of transparencies, as one did at the time, Andreessen began his self-deprecating talk: "We tried really hard not to invent anything new or solve any hard problems, which makes it easier to get something done." He then listed some of the obvious challenges his team happily ignored: "How do you search across the entire information space? I don't know. How do you know where you are going? Beats me." He then tried to describe to the audience the larger purpose of the Web, saying it is "fundamentally about communication. The applications that are going to be successful are the ones that tie together people."[49] It was a bravura performance that in a few words, we can now see, sketched out the commercial history of the Web, identifying niches in the ecosystem that would feed tech titans like Yahoo, Google, and Facebook.

Soon after he flicked off the projector, however, came the inevitable audience question: "What is the Mosaic Communications business model?" Without missing a beat, Andreessen answered: "Making money."[50] Another jokey comment, certainly, but perhaps the only one that could capture what he and Clark were thinking. Making money was the plan and, to that end, MCom would need the Web to be entertaining and useful. If no one wanted to use the Web then no one would need whatever MCom planned on selling. However, if the Web proved useful and entertaining, then businesses might pay MCom for servers to host their websites or for help running their online business. Or maybe MCom's Web site would become a valuable portal in its own right, as people new to the Web inevitably visited there to learn more about the browser that opened up the Web.

Microsoft had already demonstrated that there was money to be made from becoming the standard software for operating your computer. In Gates's days, you wanted to enforce your dominant position by preventing any copying or sharing—you withheld the program or operating system until the public came crawling, dollar bills in hand. In the two decades that separated Gates from Andreessen, however, the Web changed what qualified as an astute business strategy. A company like MCom didn't need to wield its power quite so heavy-handedly. Why send interested users away if acquiring a large audience was how a Web business ultimately hoped to make its money? In that sense, Andreessen the professional programmer had very similar priorities to Andreessen the undergraduate programmer. They both pursued the "shareware model of free distribution. . . . Get it out there and into people's hands."[51] The browser initially wasn't free, but was free on a trial basis, and the company didn't object if a user began a new trial over and over again. Gaining a large audience was so important that Andreessen was happy to suggest opportunities for entrepreneurs to explore, knowing that the browser maker should come out fine in the process as the proverbial store that sells pick axes during the gold rush.

Weeks after the San Francisco talk, MCom changed its name to Netscape Communications, and the browser's name to Navigator, in response to a complaint from the University of Illinois that it could be confused with the original Mosaic browser. The university also claimed that despite a thorough rewrite, the Navigator browser still contained code from the original version of the program that Andreessen and Bina first put together at the NCSA.[52] Andreessen is still angry about the experience, which he recounted on Twitter: "Netscape never got rights to Mosaic. We rewrote code base from scratch," he wrote. "Univ Illinois then threatened to sue us and tried to kill our business. So we sued them to stop harassment. . . . Univ of Illinois got small-$ cash payoff. Refused stock."[53] Andreessen enjoys reminding the world that had the university wisely taken stock instead of the $2.7 million Netscape paid in compensation it could have netted $5 million more.[54]

For all of his foresight in that early talk about the way the Web would grow and organize itself for commerce, Andreessen never mentioned advertising, which would become the predominant economic engine of the Internet. For the time being, Netscape's path to profitability would center on promoting online commerce. The company would sell businesses powerful servers and other tools to reach their customers, while the Netscape browser would be made commerce friendly, with new features to allow secure financial transactions. "Our goal is to get millions of copies in use quickly to really start enabling the market for lots of commercial services to come online," Andreessen said in the San Francisco talk.[55]

Despite Andreessen's initial omission, the Netscape browser would ultimately prove crucial to introducing the advertising-centric, data-collecting Web we have today. Not that this was Netscape's intention. The change in the browser that would have such long-term implications for the Web was a new snippet of code created by Montulli, who called it a "cookie." This new code was meant to fix what was seen as a flaw in Berners-Lee's original design for the Web, namely users traveled anonymously "from server to server afresh, with no reference to any previous transactions."[56] This lack of "memory" on the Web—that your past wouldn't, couldn't, accumulate—posed problems. Encounters with Web sites became "a bit like talking to someone with Alzheimer's disease," Montulli wrote.[57] You may have visited a site ten times a day, every day, but you were nonetheless considered a stranger. Unless you registered at a site, sales would have to be conducted along the time-consuming "vending machine" model[58]: You want six different candy bars? Put in the money six times and pull the lever six times. The inability to retain even the most basic information about users could have been fatal to online commerce, and Montulli devised cookies as a way for Web sites to keep tabs on their regular visitors the better to sell them things.

Montulli recalls coming up with the idea in July 1994, after meeting with an in-house team focused on supporting e-commerce. The team was planning an online shopping cart system, but the browser seemingly wouldn't allow it, since there was no way for a shopper to pick something to buy, leave it in a cart, keep shopping, add something else to the cart, and again return to shopping. Montulli wrestled with a solution for days, proudly holding the line against a proposal to give each browser a unique ID as it traveled across the Web. Such a solution was antithetical to how Montulli understood the Web—yes, a browser ID would help a Web site keep track of its customers, but it would turn every visitor into an open book, as businesses could easily pool their knowledge to create an online portrait.[59] Instead, Netscape's cookies would be built around each visit, or "session," between a particular Web site and a browser. This bit of memory would be enough to speed commerce by allowing a business to recall its customer's past preferences, credit card number, earlier selections, and the like, though that information wouldn't be carried to other Web sites.

Netscape's business customers were told of this new powerful tool for commerce in fall 1994, with the release of the first browser, and advised how best to employ it. The ordinary users of that browser—who typically were not paying customers, after all—were never told.[60] Here was a vivid example of the truth in the adage, "If you are not paying for it, you're not the customer; you're the product being sold," which was coined in 2010 by a commenter to the Web site Meta-Filter.[61] Nonetheless, Montulli and the others felt good about what they had accomplished, confident that they had the interests of users—not just businesses—at heart. Yet, in one of those painful ironies that illustrate the importance of early design decisions, the user protections that Montulli cared so much about ultimately wouldn't make a difference. Web businesses found a work-around and managed to create the kind of "cross-site tracking" he dreaded from the start.

The weakness these businesses exploited, as Montulli was forced to admit later, was that Web pages typically included embedded content from a variety of outside sources, "third parties," each of whom was able to install a cookie on a visitor's browser and keep track of where she had been and what she had done. Each visit to a Web site in actuality represented many simultaneous "sessions." Certain third parties, particularly the companies that placed digital advertisements across the Web, were ubiquitous online; an ordinary Web user could cross paths with the same "third party" site after site. Thus, it might seem that the user of a Web browser was starting a new "session" with a site, even as she was continuing a session with an advertising company that began many Web sites earlier. A business like DoubleClick, which was acquired for $3.1 billion by Google in 2008, could therefore stitch together a detailed profile of a Web surfer's online life on its own, exactly what Montulli had tried to avoid.

When, in 1996, journalists in Britain[62] and the United States informed Web surfers of the surprising news that "the Web sites you're visiting may be spying on you," there were protests over cookies and calls for the government to step in.[63] Netscape was concerned enough to ask Montulli to think of a coding change to thwart third parties. "Tracking across Web sites was certainly not what cookies were designed to do, they were designed with the opposite intention," he wrote in 2013 on his blog, explaining the predicament, "but what could be done at that point to fix the problem?" He agonized for weeks and then opted to do nothing, convinced that at this point businesses trying to profile Web users couldn't be stopped: "If third-party cookies were disabled, ad companies would use another mechanism to accomplish the same thing, and that mechanism would not have the same level of visibility and control as cookies."[64]

When Microsoft introduced a browser to compete with Netscape, there never was a question about whether it would have cookies too. The gains from cookies were tangible, the loss of privacy less so. Montulli's warning about how the tracking would only get worse without cookies proved correct. Two decades later, Facebook, for example, has access to so much more information about its users than mere browsing history—what they like and dislike, whom they communicate with, their relationship status, what articles they click on, what articles they read to the end, even—and the mechanism to limit what it retains is much less visible.

Despite the many improvements Netscape introduced to assist in Web commerce, the company itself hadn't achieved commercial success. Past practice on Wall Street was for a company to hold off on an initial public offering of stock until there had been at least three consecutive profitable quarters; in the summer of 1995, Netscape was still waiting for its first. But Clark argued that Netscape should play by its own rules—after all, no other company had experienced the kind of viral growth that Netscape had, approaching 90 percent of the browser market. If Clark's experience in Silicon Valley had taught him anything it was to take the money when it's sitting there. "I wanted us to go public, because I thought it'd be good for us from a P.R. standpoint, and I did go into this thing to make money, so I was looking for a reward as well," he said.[65]

The day of the IPO, August 9, 1995, appeared to be perfectly timed, producing a mania that surprised even Clark. Shares spiked to nearly three times the opening price, from $28 to $75, before settling at $58. By the end of that day's trading, Clark's stake in Netscape was worth $663 million, a fact he recalled a little later when he needed to come up with a tail number for an airplane he bought. "I told them to use 663, because that meant something to me."[66] Andreessen clocked in with around $60 million. Jimmy Wales, the cofounder of Wikipedia, was one of a number of aspiring Internet entrepreneurs who considered 8/9/95 a life-changing date. Wales had already dropped out of graduate school to become a futures and options trader in Chicago, but "when Netscape went public and it was worth more than $2 billion on the first day," he recalled, "it clicked in my mind that something big was happening on the Internet."[67]

Shares of Netscape stock never eclipsed their peak of $171 in December, and from such heights, seemingly the company had only one direction to go. The frenzy leading to the IPO made Microsoft finally take notice of the Web, after Bill Gates had dismissed it as so much hype.[68] In a bit of painful payback for Andreessen, a version of Mosaic licensed by the University of Illinois to a local software company, Spyglass, helped Microsoft quickly challenge the Netscape browser. Working from Spyglass's browser, Microsoft released its first version of Internet Explorer the same month as the Netscape IPO, followed in October 1995 by the release of the beta version of the much-improved Internet Explorer 2.0, which was available free for surfers and businesses alike. New versions quickly followed, and Microsoft aggressively promoted them. A turning point came two years later, in October 1997, when the company released Internet Explorer 4.0, which was tightly bundled with the Windows operating system. At that point, Netscape was still roughly twice as popular—65 percent of the market versus 32 for Internet Explorer.[69] In November 1998, however, the tide was turning, and a declining Netscape was bought by AOL in a stock swap valued at the time at $4.2 billion.[70]

Andreessen's wild run from Mosaic to Netscape to AOL thoroughly transformed the computing world. Microsoft may have succeeded in taking down Netscape—with Internet Explorer, for example, reaching a peak market share well above 90 percent—but that success took a toll. The company's win-at-all-cost approach to "the browser wars" produced important evidence when the U.S. government filed an anti-trust lawsuit against Microsoft in 1998, particularly testimony alleging that Microsoft had conspired to cut off Netscape's "air supply." After Microsoft settled that lawsuit in 2001, Microsoft didn't have the same strut or chokehold on how the public used computers. In a fitting final twist to the Netscape-Microsoft fight, the code for Netscape Navigator was released as free software—that is, free to be shared and improved upon by whoever acquired a copy. That code became the basis of the Firefox browser (original name Phoenix because it rose from Navigator's ashes), which helped chip away at Internet Explorer's dominant market share. Today, Firefox lives on as part of a nonprofit project supported by a community of programmers who were motivated to push back against Microsoft.

Andreessen is a dual figure, the hacker-entrepreneur. At the same time that the Netscape team helped make the Web commerce friendly, it also helped install anarchic hacker values onto the Web. For example, on the Web, as in the artificial intelligence lab, little deference would be given to authority: anyone can publish online no matter his age or experience. Your work speaks for itself. The Web also adopted the hackers' belief that information should be free to circulate: music files, newspaper articles, movies, and software all bouncing from computer to computer, unrestrained by duplication costs and seemingly one step ahead of the authorities. There was also a similar consensus that freedom of speech should trump all other concerns: the Web would be beyond the reach of "politically correct" censors declaring some comments as too hateful or cruel or obscene to appear. A more society-focused vision of the Web lost out, although it has been kept alive on the margins, often by European governments who try to prop up traditional newspapers as a stabilizing force and where some topics, like far-right political parties, are barred from appearing online. Some European governments have gone so far as to resist computers' unrivaled memory skills, promoting a right to be forgotten so that an individual can insist that material be taken off the Web if it is old and embarrassing.

In the early 1990s, administrators at the NCSA had briefly suggested that the Mosaic browser warn users if a Web site might not be suitable for children. The problem arose from a link on the What's New page Andreessen maintained for the Mosaic homepage, back when a single person could actually try to keep up with what was new on the Web! A child of a lab employee had clicked on the link and was sent to an arts site with a prominent display of a nude sculpture. Administrators asked Andreessen to come up with a fix. It was the Stanford censorship case all over again, and the hackers' loyalties hadn't shifted. They knew that censorship was stupid and antithetical to the Web, and Andreessen offered a suitably stupid proposal. Let's have a box appear before a user reaches any new Web site, he suggested, with the following warning: "ARE YOU SURE YOU WANT TO TAKE THIS CRAZY STEP AND KEEP SURFING?" The administrators decided to pass.[71]

One clear articulation of how hackers' anti-authoritarian views were shaping the Web appeared in 1996 as "A Declaration of the Independence of Cyberspace," written by John Perry Barlow, the libertarian cofounder of the influential digital rights group the Electronic Frontier Foundation. No government, Barlow declared, had the authority to limit the freedoms inherent to cyberspace. "Governments derive their just powers from the consent of the governed," he announced. "You have neither solicited nor received ours. We did not invite you. You do not know us, nor do you know our world. Cyberspace does not lie within your borders." More broadly, Barlow was arguing that nothing from the offline world—traditional rules, institutions, and codes of behavior, even history itself—carried any weight in cyberspace, which was "a world that all may enter without privilege or prejudice accorded by race, economic power, military force, or station of birth." Having ditched America's living history of racism in less than a sentence, and ignored the misogyny outright, Barlow was then free to demand the familiar absolutist line about online speech. "Anyone, anywhere," he wrote, "may express his or her beliefs, no matter how singular, without fear of being coerced into silence or conformity."[72]

The declaration is a political statement about as nuanced and considered as the hand-scrawled "Keep Out" sign that a teenager tapes on his door. Nonetheless, it accurately describes much of the Web today—the hostility to authority and rules or regulations of any kind; the privileging of freedom over empathy; the fantasy that the Internet is immune to the pull of history. Barlow certainly drew inspiration from the early hackers as he wrote his declaration, but the radical ideology promoted by the Know-It-Alls over the last twenty years involves so much more than the hackers' desire to be left alone. Other troubling aspects of Silicon Valley values described in this book—blind faith in the power of markets to do good all the time, trafficking in people's private information as a commodity, acquiring obscene personal wealth and pursuing economic and social disruption for their own sake with no thought to the human cost—have nothing to do with the hacker ethic. In fact, McCarthy and the other early hackers were critical of those who saw the computer revolution as a path to personal wealth or its close cousin, personal power. Even Lou Montulli recalls being taken aback by the promise of a quick fortune from Jim Clark. "He filled our heads with giant numbers of how we were going to make riches and be the most important people on the planet," Montulli recalled,[73] which conflicted with his own "sort of Marxist" belief that "you couldn't make more than a million dollars honestly."[74]

After a decade of trying to replicate the success of Netscape at other ventures, Andreessen in 2009 found his calling as a Silicon Valley venture capitalist. He promotes disruptive capitalism among a new generation of hacker-entrepreneurs who, as he memorably put it in the title of a *Wall Street Journal* op-ed, create software that "is eating the world."[75] Certainly he was better suited to be an investor than to run a company of his own. To start, his misanthropic personality, a liability in a manager, is seen as an asset in an investor.[76] His partner, Ben Horowitz, explains: "If you say to Marc, 'Don't bite somebody's fucking head off!', that would be wrong. Because a lot of his value, when you're making giant decisions for huge amounts of money, is saying, 'Why aren't you fucking considering *this* and *this* and *this*?'"[77]

Think back, too, to Andreessen's 1994 talk to aspiring online entrepreneurs: when it came to how the Web would adapt to become more business friendly he was brimming with ideas, but as to the details of the business plan for his own company, he offered the generic "make money." In his role as a VC, he is expected to survey the economy, weighing in as a public intellectual and pointing to broad trends. Back in 2003, when Andreessen was still an entrepreneur, he scoffed at a reporter who wondered if he kept a personal blog. "No," he responded,

"I have a day job. I don't have the time or ego need."[78] More recently, however, he has been inclined to marvel at the influence of his Twitter feed, where he had posted more than 100,000 times and acquired an audience of half a million followers. "Reporters are obsessed with it," he bragged to a reporter. "It's like a tube and I have loudspeakers installed in every reporting cubicle around the world." Andreessen was using Twitter as a Trumpian bullhorn for his ideas of software-led disruption before Donald Trump was a serious enough figure to communicate effectively in 140-character bursts.

On Twitter, Andreessen's praise of disruption has been exclusively economic, not political or social. For example, in a Twitter "essay" from 2014, Andreessen praised the slow and deliberate steps America has made in opening up its political system to all citizens. Don't dwell on all the improvements yet to be enacted, he advised, but instead think about how far we've come. "Common thing one hears in US is 'Political system broken; Founding Fathers never intended politics to be dominated by moneyed interests.' But in 1776, voting 'restricted to property owners—most of whom are white male Protestants over the age of 21.' In 1789, George Washington was elected president. 'Only 6% of the population can vote,'" he wrote, adding, "We have far broader-based voting and political participation today than ever before, due to hard work by many activists over 200 years. And we're still by no means perfect; lots of progress yet to be made. But we're leaps and bounds ahead of 50-100-150-200 years ago."[79]

Imagine Andreessen's reaction to someone who made a similar argument concerning the economic disruption caused by Internet companies—you know, think how far we have already come, let's not act too hastily. Well, we don't have to imagine, actually. In another Twitter essay, Andreessen argued that technological progress has benefited the poor much more than the rich—an observation he insists "flows from basic economics." Therefore, he writes, "Opposing tech innovation is punishing the poor by slowing the process by which they get things previously only affordable to the rich."[80] To recommend patience in implementing technical changes is simply immoral. What's the difference? Well, one difference is the power relationship. In the case of the disruptive democratic politics that Andreessen appears leery of, members of the public are being given greater control over their lives at the expense of an elite; in the case of disruptive technologies, an elite is driving the change.

In September 2016, just when Trump was deploying Twitter to strike out at his foes and communicate without speaking to the press, Andreessen stopped posting to Twitter, not long after having to apologize for posts that praised British colonialism in India as superior to democracy in providing for the poor.[81] In a clever work-around, however, Andreessen has remained active on Twitter by "liking" as many as forty posts a day written by others. That way he seemingly expresses his opinion—and continues tangling with Web idealists—without bearing ultimate authorial responsibility for what has been said.

The first idealist Andreessen ever tangled with, of course, was Berners-Lee. In his memoir, Berners-Lee is quick to reject the anti-capitalist label, denying that he thinks the Web should be treated as some hallowed space, "where we must remove our shoes, eat only fallen fruit and eschew commercialization."[82] But he also clearly isn't comfortable with how it has been twisted to generate runaway profits. At one point in this same memoir, Berners-Lee pauses to answer why he never tried to amass a fortune from his ideas, even as so many other key figures in the Web's development did. "What is maddening is the terrible notion that a person's value depends on how important and financially successful they are, and that that is measured in terms of money," he writes. "To use net worth as a criterion by which to judge people is to set our children's sights on cash rather than on things that will actually make them happy."[83]

Berners-Lee tells a story about a technical breakthrough in the development of the Web that occurred on Christmas 1990. That day, his computer at CERN for the first time used a primitive browser/editor to communicate with a server hosting the Web's first URL, info.cern.ch. The Web worked! Even so, Berners-Lee writes that he "wasn't that keyed up about it," because he and his wife were expecting their first child and, "As amazing as it would be to see the Web develop, it would never compare to seeing the development of our child."[84] Even at this late date, we would do well to try to restore the human-scale perspective and idealism Berners-Lee brought to the Internet project from the start. The lesson the Know-It-Alls took from those early years, however, was to grow big and grow fast.

5 . JEFF BEZOS

"When it's tough, will you give up,
or will you be relentless?"

I n his talk in 1994 to wannabe Web entrepreneurs, Marc Andreessen imagined a system of e-commerce that resembled small-town
life. "The corner pizza store in a box" is how Andreessen pitched
Netscape's product that included a server and software. "You put up
your corner pizza store, your flower shop, your bookstore. I really
believe that this is where things are heading. It is going to be interesting when it arrives."[1] This system wouldn't be quite as decentralized
as it looked, however. Netscape, as the provider of the browser, server,
and related software, would become the beneficiary of the network
effect that comes from holding the dominant position online. Netscape
products would be what everyone used, and therefore what millions of
new arrivals would choose to use when they joined the Internet party.
This was the dream, at any rate, albeit a short-lived one. The next year,
1995, the Web took a sharp turn toward centralized online marketplaces as Craigslist, eBay, and most notably Amazon got their first
taste of how being popular online only made you more popular.[2]

Craigslist and eBay both grew organically, almost by accident. Craig
Newmark, a former IBM programmer living in the Bay Area, began
maintaining and distributing his "list" of local events and job open-
ings in early 1995. He sent the listings as a mass email to friends,

at first just ten to twelve people. Word spread quickly, and to meet the growing demand Newmark soon professionalized his system for sending the emails. In early 1996, Craigslist—its nickname from the start—migrated to the Web. By the end of 1997, Craigslist was getting about a million page views a month and Microsoft approached him about running banner ads on the site. Newmark recalls turning down the offer, knowing in doing so he had "stepped away from a huge amount of money."[3] Though Craigslist was a for-profit enterprise (except for a short stint when it switched to nonprofit status), Newmark wasn't in a particular hurry to grow Craigslist or to cash in. He was making a good living as a programming contractor and his priority was Craigslist's users, not making a bunch of money.

To this day, the site has been slow to assess fees of any kind, limiting them to certain job and real estate listings as it has steadily added cities and countries to its service. In 2000, Craigslist expanded from San Francisco to nine other United States cities, including Boston, New York, and Chicago; the next year, Vancouver became the first city outside of the United States. In 2003, the London site launched, the first outside of the Americas.[4] "Remember," Newmark said, "in the conventional sense, we were never a startup. In the conventional sense, a startup is a company, maybe with great ideas, that becomes a serious corporation. It usually takes serious investment, has a strategy, and they want to make a lot of money. We've done something very different."[5] As Craigslist's chief executive, Jim Buckmaster, once explained to a room full of equity analysts and fund managers, Craigslist has no business development team and politely declines any offer that shows up in its inbox.[6] Newmark describes himself as the company's founder and customer service representative.

EBay was born a bit later in 1995 as AuctionWeb, the project of another programmer in the Bay Area, Pierre Omidyar, who at first hosted the online marketplace on his personal Web site.[7] His plan for publicizing AuctionWeb included posting notices on Usenet groups and getting listed on the What's New page at the supercomputing center at the University of Illinois, which was created by Marc Andreessen when he was still an undergraduate. Through such informal channels, Omidyar reached enough Web users to have hosted thousands of free auctions with tens of thousands of bids by the end of 1995. Word of mouth did the rest. His personal Internet service provider noticed the spike in activity and raised his monthly fee from $30 a month to $250—the business rate—even though the site wasn't yet operating as a business. "That's when I said, 'You know, this is kind of a fun hobby, but $250 a month is a lot of money,'" Omidyar recalled.[8]

Omidyar started charging a fee that was a percent of the final sale price, a system well designed for ensuring that you had the resources to grow as fast as your audience would take you. The project's first employee and first president came the following year; in June 1997, a Silicon Valley firm, Benchmark Capital, invested $6.7 million for what became approximately a quarter of the company. Omidyar deposited the check, but didn't touch the money, he said. More than its capital, Benchmark was providing prestige, connections in hiring, branding, and marketing advice[9]; in September, AuctionWeb was officially renamed eBay.[10] After eBay had its successful IPO in 1998,[11] many called the Benchmark deal "the best-performing Silicon Valley investment ever."[12]

Amazon.com offers a stark contrast to these hacker-founded marketplaces, which bloomed naturally, one as a popular, profit-agnostic Web site, the other as a Silicon Valley–endorsed tech titan. Amazon, too, started small in July 1995 as an eager-beaver online bookseller, but the intention was always to become the one-stop shop for all online commerce. Nothing accidental about its path to greatness. First books, then everything else. The plan was hatched by a young Wall Street analyst, Jeff Bezos, while he worked for D. E. Shaw & Co., an elite "quantitative" hedge fund founded by David E. Shaw, who was a veteran of McCarthy's artificial intelligence lab at Stanford in the 1970s. "The idea," which Shaw says occurred to him and Bezos, "was always that someone would be allowed to make a profit as an intermediary. The key question is, 'Who will get to be that middleman?'"[13]

At Stanford, Shaw had researched how to improve the underlying architecture of thinking computers, periodically leaving the lab to start high-tech ventures. He earned his computer science PhD in 1980 after he "was eventually coerced into going out of business by his thesis supervisor."[14] Shaw's years at the lab were spent "living in the future," he recalled, with robots wandering the grounds and music streaming from the lab's PDP-10 computer. The future appeared well confined within those walls, however. In a message to his former colleagues, Shaw wrote: "Although I'm sure that some of you foresaw more than I did, I can't remember sharing a collective vision, for example, of how the Arpanet we took for granted then might someday turn into anything remotely resembling the Internet we take for granted now."[15] For all its obvious innovation, the Stanford lab of that era was inward looking, skeptical of entrepreneurism, and proudly peculiar, like its leader, John McCarthy. The message was clear: to make a mark in society and put some cash in one's pocket, best to leave the lab and its quixotic pursuit of artificial intelligence and look for a market to disrupt.

After earning his degree, Shaw initially continued his computer science research at Columbia, constructing the supercomputers that he had designed while a Stanford graduate student. In time, he drew up a business plan based on that research, which is how Shaw first crossed paths with Morgan Stanley investment bankers. Shaw's start-up idea fell through, but the bankers' pitch to him to bring his computer skills to investing found a receptive audience. "I couldn't help wondering whether state of the art methods that were being explored in academia could be used to discover the other investment opportunities that weren't visible to the human eye," he recalled.[16] In 1986, Shaw joined Morgan Stanley, where he applied his high-speed computers and sophisticated algorithms to financial markets, earning six times his assistant professor's salary.[17] When things were going well, Shaw's computers and algorithms could find assets that hadn't reached a stable, global price, allowing Morgan Stanley to buy that asset where it was undervalued and sell it where it was overvalued. In other words, artificial intelligence could discover a sure thing, the dream of everyone who plays the horses or the stock market.

Though Shaw's computers and accompanying algorithms produced a handsome return on investment, they didn't represent an important breakthrough about the nature of intelligence. Like the machines of that era that ran chess programs that could beat even elite players, Shaw's investing algorithms may have been much smarter than their predecessors, and more capable than people, but not because they were better at independent thought. They benefited from faster computers that could process more information, combined with the fact that there was so much more financial information to be processed from an increasingly global economy. However, there was one obvious difference between chess and investing—the extent of the prize money. After a couple of years' success at Morgan Stanley, Shaw decided to strike out on his own and was able to attract $28 million in capital.

D. E. Shaw & Co., which began trading in 1989, moved into offices in Manhattan's Flatiron district, above a famous Marxist bookstore.[18] The dress code was straight out of the computer lab, and even in those pre-Web days the firm's employees were equipped like the members of a lab—Sun SPARCstations with Internet access, which could be used for email and analysis.[19] At this computer lab/investment firm, however, there was a single, overarching research goal: "To look at the intersection of computers and capital and find as many interesting and profitable things to do in that intersection as we can."[20]

Hiring at the firm was treated with academic rigor, focused exclusively on finding the biggest brains with the best degrees. Shaw didn't wait to hear from interested candidates and instead sent unsolicited letters to top students that explained, "We approach our recruiting in unapologetically elitist fashion."[21] The interviews at Shaw were grueling and meant to explore how a candidate thinks by asking gnarly questions like, How many gas stations are there in the United States? This method of detecting genius by posing riddles has been popular in Silicon Valley at least since the 1950s when the transistor pioneer William Shockley began hiring for his influential start-up, Shockley Semiconductors. Shockley grew up in Palo Alto, where he knew the Termans. To his lifelong shame, Shockley was tested for Lewis Terman's study of gifted children and came up short. Shockley nonetheless relied on tests of all kind—personality, intelligence, lie detector—to assess job candidates as well as to keep tabs on current employees. The more about a person that could be converted into a number, from Shockley's perspective, the better.

For Shockley, however, part of the point was that no candidate should score higher than he did. One time, a young physicist, Jim Gibbons, was asked during his interview to figure out how many tennis matches were required to settle a singles elimination tournament with 127 players. Shockley started his stopwatch, expecting to see a bunch of calculations, but in a barely a moment, Gibbons gave his answer: 126. The candidate explained his logic: there is only one winner, with

126 others eliminated; a match is needed to eliminate a player; therefore, 126 matches to eliminate 126 players. Shockley replied in fury, "That's how I'd do it!" He wanted to know if Gibbons had been tipped off. Only when the next question stumped Gibbons was peace restored to the hiring process, though in truth, Shockley never found peace from tests.[22] He spent his later years in the 1970s and 1980s as a reviled figure, an emeritus professor of engineering at Stanford who would travel the country to promote his claims that blacks were intellectually inferior and advocating voluntary sterilization programs for mothers with low IQs.[23] Without any personal sense of irony, the Nobel prize–winning Shockley would cite results from a test that found him not-quite-genius material to make his case for group differences in intelligence.[24]

When Jeff Bezos arrived for his interview at the Shaw offices, he was already prepared to leave Wall Street to somehow start his own business, but he discovered a soul mate in David Shaw.[25] Shaw soon promoted Bezos to cover the Internet for the firm. In no time, Bezos and Shaw became convinced of the Web's nearly boundless commercial potential, which was reflected in the unprecedented burst in traffic in 1993 as the Mosaic browser was quickly being adopted. The question for Bezos became whether to start a Web commerce site as part of Shaw's company or to begin his own. In the spring of 1994, Bezos scheduled a heart-to-heart with his mentor, who "took me on a long walk in Central Park, listened carefully to me, and finally said, 'That sounds like a really good idea, but it would be an even better idea for someone who didn't already have a good job.'" Bezos considered the point for forty-eight hours, and decided to leave Shaw and his 1994 bonus. Looking back on that crucial meeting as part of a speech to Princeton graduates, Bezos concluded that life was best considered as a series of choices: "Will you follow dogma, or will you be original? . . . Will you play it safe, or will you be a little bit swashbuckling? When it's tough, will you give up, or will you be relentless? Will you be a cynic, or will you be a builder?"[26]

Bezos was a builder, clearly, and was prepared to play the long game with his Web site, which wouldn't appear online for nearly a year. At thirty years old and after nearly a decade on Wall Street, Bezos wasn't some young hacker who stumbled on a great Web idea and then would shyly go about inquiring how to produce a business plan. He was a young investor who enlisted hackers to carry out the business plan he had mapped out with spreadsheet projections and a general's sweep. Books were the perfect beachhead for an aspiring e-commerce giant trying to build a loyal customer base, he concluded. A book, unlike a razor, a pizza, or a blender, could inspire. People loved books. People gave books as presents. "We've all had books that have changed our lives," Bezos said early on, "and part of what we're doing at Amazon .com is trying to help you find and discover that next one."[27]

In the year between deciding to start his own e-commerce Web site and when that site would go live, Bezos raised capital from friends and family, hired employees, and prepped and tested the Web site. Bezos was proficient enough technically to know that he needed to hire professionals to design a site that could win an audience and accommodate rapid growth. His favorite to do that initial build would have been Jeff Holden, who had recently graduated with undergraduate and master's degrees in computer science from the University of Illinois, Urbana-Champaign. An alumni newsletter traced Holden's path out of Illinois: "As a good, hard-working student, Holden caught the attention of a recruiter for D. E. Shaw, the investment powerhouse. Holden had never given investment banking a thought, but not one to say no to something he didn't understand, he decided to check out the company. Mesmerized by New York City, the amazing people at Shaw, and the physical presence of Shaw's corporate headquarters, which he described as 'a super sexy, totally artistic view of the office space,' he took the job." At D. E. Shaw, Holden and Bezos worked together, and when Holden heard his colleague would be starting an e-commerce site he was eager to join. Unfortunately, Bezos couldn't act until two years had passed as part of a nonpoaching agreement with Shaw.[28]

Holden would join Bezos "two years and four seconds" later, but in the interim Bezos signed up a West Coast hacker named Shel Kaphan, who built the first version of the Amazon Web site. Kaphan came with less pedigree and was older—he was self-taught and had been hacking since the 1970s.[29] Kaphan felt he had missed out on the earlier computer revolution, in part because he lacked a strong business sense. Seemingly, nothing had changed. Even before speaking with Bezos,

Kaphan was thinking about how the Web could help people find the books they were interested in. But his idea was to hack the clunky notecard system used in libraries by creating a clickable digital version of *Books in Print*. "I wasn't thinking about it in the context of selling books," he explained, "but I was thinking, 'Man, I hate going to the library and ruffling through those card catalogues and trying to find that thing that I'm looking for.'"[30]

Bezos, however, was only thinking of selling books. Creating an online index along the lines Kaphan described was crucial to Amazon.com's impressive claim to give its customers access to "a million titles." As a Web start-up, the company wouldn't actually stock copies of those million titles, but neither did the biggest bookstore, for that matter. Book buyers were accustomed to learning that a book they wanted wasn't in stock and would be delivered later. Bezos was introduced to Kaphan through a friend of a friend who worked with Bezos at D. E. Shaw. After a few meetings in Santa Cruz, California, Kaphan agreed to build the Web site for a company then still known as Cadabra, as in the magical abracadabra. The meetings were in late spring 1994, and Kaphan's participation was contingent on where Bezos decided to put down roots. At the time, the choice was between Seattle and somewhere in Nevada, with California out of the picture, Kaphan recalled, because of its sales tax. When Bezos chose Seattle over Nevada, Kaphan was in—employee No. 1. He arrived in October 1994 and went about acquiring the computers, software, and expertise to build the site.[31]

In a month, Kaphan was joined by another programmer and in spring 1995, Amazon had a "friends and family" soft launch. In July, the site had its hard launch as Amazon.com. Among the many attributes Bezos liked about the new name—besides conveying the size and breadth of the world's largest river—was that it began with an *A*, and would be at the top of the What's Cool, What's New pages on the Netscape home page, the heir to that first list on the Mosaic home page. Barely a week after launching, Yahoo, an up-and-coming Web index created on the Stanford campus by two engineering graduate students, offered to feature Amazon.com on the site. Kaphan worried if Amazon.com could handle the surge in traffic, but Bezos said yes, setting the stage for a turbo-charged start.

Within its first month, Amazon.com had sold books to residents of all fifty states and of forty-five countries.[32] Even so, the book distribution system posed an early challenge because distributors had a rule that they had to ship a minimum of ten copies at a time. At the start, Amazon may only have needed a single book from a certain distributor. Bezos discovered a hack: the distributors required that ten books be *ordered* at a time, not that ten books be *delivered* to a client. "So we found an obscure book about lichens that they had in their system but was out of stock," he explained. "We began ordering the one book we wanted and nine copies of the lichen book. They would ship out the book we needed and a note that said, 'Sorry, but we're out of the lichen book.'"[33] That would not be a lingering problem. The site's traffic was doubling every quarter as Amazon expanded its purview, first by adding music and videos, which necessitated changing its motto from "Earth's Biggest Bookstore" to "Books, Music and More."

After two years, Bezos was free to hire D. E. Shaw employees, and Kaphan found himself being pushed aside. "The company started growing and we started attracting zillions of MBAs," he recalled. "It just stopped being fun."[34] He remained at the company for five years, so that all his stock could vest, but he was well out of the loop. Kaphan bears a grudge toward Bezos about how he was treated, even if that intense period of work left him very wealthy. He wonders if his work benefited society. "At this point, I don't know," he told an interviewer. "When I look at technology these days, I see that it's either doing something to connect people or it's doing something that isolates people." Amazon, in particular, he says, "is more on the isolating people side. Everything caters to convenience so much that you don't even have to get out of bed to take care of your day-to-day business. To me, that's a step too far."[35]

In early 1999, Amazon bought a controlling share in Drugstore.com and began a service called "Shop the Web," which added still more products for online purchase. Bezos was frequently raising money, which allowed Amazon to propose "megadeals" with the largest Web portals like AOL, Yahoo, MSN, and Excite; Amazon would pay tens of millions of dollars to be the exclusive bookseller and to have its results show up on searches. This was inorganic growth, to be sure, but Bezos considered growth too important to leave to chance. Around the time of the Drugstore.com acquisition, a journalist tried to explain Bezos's drive to expand before even returning a profit. "While his beleaguered physical rivals are mired in the present, where they have to attend to sticky details like making money, the specifics of Amazon. com's ultimate form remain forever elusive, a lovely shimmering at the edge of the horizon," he wrote. "In this way Amazon.com truly is a virtual company, existing only in the imagination."[36]

Amazon has continued to expand to encompass all aspects of commerce, including an increasingly lucrative business selling cloud-based computing. The company now has a dominant share of online retail sales and has become for all intents and purposes the "middleman" for online commerce that Bezos and Shaw imagined, shaping our economy and our lives. In 2015, Amazon surpassed $100 billion in sales and registered 300 million customers; analysts estimate that more than half of any growth in e-commerce will go directly to Amazon.[37] Upon achieving that $100 billion milestone, Bezos was prompted to look back twenty years to when he was "driving the packages to the post office myself and hoping we might one day afford a forklift." Even as he celebrated how far Amazon had come, he noted that, "measured by the dynamism we see everywhere in the marketplace and by the ever-expanding opportunities we see to invent on behalf of customers, it feels every bit like Day 1."[38]

There is something rejuvenating about approaching every day as if it were your first. But something unnatural, too. The workplace pressure never ceases—each day you must return to the marketplace to have the public render its verdict. Again. Again. And again. The Web-based marketplace keeps everybody productive and on their toes by instantaneously shuttling information between customers and businesses. If Amazon somehow isn't attentive to its customers, the dollars quickly go elsewhere. The tricky issue for Bezos has been to look beyond how this system serves customers to consider how the system serves Amazon's workers.

In 2015, the *New York Times* published a detailed report that portrayed Amazon as a ruthless employer on behalf of its customers the company's white-collar workers were crying at their desk from the stress and pressure.[39] And this account didn't begin to address the treatment of blue-collar workers in Amazon's warehouses around the world, which, depending on where on the globe they are and depending on the season, can be unbearably cold or unbearably hot.[40] Bezos expressed disbelief at the *Times* account—he was confident, he wrote in a note to employees, that he would never, could never, lead the kind of "soulless, dystopian workplace where no fun is had and no laughter heard," as described in the *Times*.

The market simply wouldn't allow it, he explained to his employees. "I don't think any company adopting the approach portrayed could survive, much less thrive, in today's highly competitive tech hiring market," he wrote. "The people we hire here are the best of the best. You are recruited every day by other world-class companies, and you can work anywhere you want."[41] This response appeared to be willfully ignorant of the pressures on employees to fight through unpleasant work experiences, to resist switching jobs. But more so, Bezos's response revealed his unalloyed libertarian faith in markets as the path to a moral workplace and a moral society. He had convinced himself that the market would somehow protect workers.

6. SERGEY BRIN AND LARRY PAGE

"It was like, wow, maybe we really
should start a company now"

In a matter of a couple of years, the anarchic Web of Tim Berners-Lee was downright business friendly as Web sites took advantage of tools to keep tabs on their visitors and accept secure payments. These were important first steps in imposing some order, but they didn't begin to address the fact that the Web itself was a shambles . . . by design. A growing, decentralized, seemingly chaotic mess, the Web had resisted attempts to be thoroughly mapped, either by people or by computers. By November 1997, there were an estimated hundreds of millions of Web pages,[1] yet as Marc Andreessen observed early on, the Netscape development team simply skipped the hard navigation problems in order to get a browser to market quickly. "How do you search across the entire information space?" Andreessen asked rhetorically. "I don't know. How do you know where you are going? Beats me. The sort of surprising thing is we ended up with something useful anyway."[2] To Sergey Brin and Larry Page, Stanford PhD candidates in computer science who met in 1995, such questions weren't so disposable. If users didn't have the means to discover what was on the Web, then it may as well not be there. How, Brin and Page asked, could the Web ever meet its potential as a vital new public resource?

At its start, the Web was just small enough to be approached on a personal level. Its contents could be organized by indexers, modern-day equivalent of librarians, who kept up with what new sites had arrived on the scene and discovered interesting themes and connections among sites. There were potential flaws with this system, which would be particularly obvious to computer scientists like Brin and Page. "Human maintained lists cover popular topics effectively," they wrote, "but are subjective, expensive to build and maintain, slow to improve, and cannot cover all esoteric topics."[3] In Silicon Valley jargon, the problem with human indexers was that they couldn't "scale" along with the Web. People were too expensive for the task, especially considering how little processing power they came with.

In a biting account of her time as an English major working at Facebook, Katherine Losse explained the importance of scalability to Silicon Valley. "Things were either scalable, which meant they could help the site grow fast indefinitely, or unscalable, which meant that the offending feature had to be quickly excised or cancelled, because it would not lead to great, automated speed and size," she wrote. "Unscalable usually meant something, like personal contact with customers, that couldn't be automated, a dim reminder of the pre-industrial era, of human labor that couldn't be programmed away."[4] The answer to meaningful Web search would have to come through computers, Brin and Page were convinced, not squads of librarians. Up to that point, however, computers had been found deficient, too.

One notable attempt to search the Web with computers was Alta-Vista from the Digital Equipment Corporation. AltaVista began December 15, 1995, with an impressive 16 million Web pages indexed by its computers, which could be searched through to answer a query.[5] AltaVista wasn't very intelligent, however. It focused on keywords in Web pages, seeing how often and with what prominence those words appeared when deciding what results to return for a query. What keyword-based searches typically produced, in Brin and Page's professional opinion, were "junk results" that "wash out any results that a user is interested in."[6] Not only was the technique unsophisticated, it was ripe for manipulation by unscrupulous Web sites, which added potential keywords to their text willy-nilly. If one mention of the word *car* on a page helped it show up in the results for a search about automobiles, the thinking went, then surely twenty *cars* would be even better.

Seeing a computer act so stupidly was an affront to Brin and Page. The tools of artificial intelligence, they believed, would allow a computer to understand what was on the Web and help guide users to exactly what they were searching for. Their insight was to use the clickable links between Web pages, rather than keywords. Links, it turned out, conveyed much more information about a site than parsing which words appeared most frequently. Each link announced, in essence, I've been to this other Web site and found something helpful there. In addition, the text underlined by the link explained what in particular was so helpful—one person's attempt to map what she had discovered about the Web. Perhaps those millions of Web links and descriptions, interpreted by a clever algorithm, were all that was needed to produce an effective, automated online catalog. The best part was that those millions of links were just sitting there for the taking. Thanks to the Web's open design, Brin and Page, or anyone for that matter, in most cases could freely collect as much of the material on the Web as they wanted, whether individual links or copies of entire Web sites.

Brin and Page viewed this project in strictly academic, philanthropic terms at the start. They were computer scientists planning to follow in the footsteps of artificial intelligence pioneers like John McCarthy, who was a living legend to Brin and Page, walking the same halls of the Stanford computer science department. Their search engine would be tackling familiar AI questions—how does a computer receive information about the world and can it be trained to interrogate the information it receives to make real-world judgments? Had all gone according to plan, the two young academics would have created their uncannily accurate search engine, published the results in an important academic paper, earned their PhDs, and become professors. At the same time, the search engine they developed—at first called Back-Rub and later Google, in reference to the absurdly large number called a googol—would remain noncommercial and freely available to the public through Stanford.

When Brin and Page reversed course, however, and reluctantly dropped out of graduate school to run Google as a business, they would be sharing with the world a different discovery—an important new technique for profiting from the Web. By this time, there were companies like Netscape and Microsoft, among others, that profited from creating the software used online. There also were companies like Amazon and eBay, soon to be joined by PayPal, which intended to take a cut from online commerce. Google, however, would pave the way as a big business that extracted value from the ordinary material folks posted online while they were going about their digital lives.

Executing Brin and Page's innovative ideas for building a high-functioning search engine posed a number of challenges to surmount. The computers that "crawled" the Web, making copies of pages and collecting the links, were expensive, but Brin and Page were not above cadging what they needed by hanging out on the loading docks to learn who in the computer science department might be getting free equipment he didn't need.[7] Brin and Page then had to devise efficient methods for searching through the oodles of information they were collecting to isolate Web links. The last challenge: how exactly should the algorithm interpret those links?

Page and Brin, the sons of professors, used a technique that was similar to how the academic world evaluates a scholar's reputation. Say there were two scholars you wanted to compare; each has articles you haven't read or that were beyond your understanding. A natural approach would be to seek the opinions of experts in the field. You might even automate the process by sending out a survey to many scholars and then counting how often they had cited each candidate's research in their own. The higher the total, the more important the scholar, you might conclude. If you wanted to get fancy you might devise some formula—an algorithm—that didn't merely count the number of citations but somehow simultaneously factored in the importance of the scholars whose citations you were using. After all, if you were trying to assess the quality of somebody's work, you would do well to assess the quality of the assessors, too.[8] Think back to Frederick Terman's time as Stanford's provost, when he was determined to raise the university's stature in comparison to the top schools back east. Armed with a slide rule and similar kinds of formulas, Terman had no qualms about determining the precise numerical value of scholars far from his area of expertise—say, a proposed new English professor or the entire history department faculty. With these tabulations at the ready, Terman would then confidently rank the hiring of this person compared to the hiring of that one.

In many ways, a search engine faced a similar predicament to Provost Terman's. It, too, was trying to assess the quality of something it didn't or couldn't personally "understand"—in the search engine's case, a random Web page written by a human being. Still, Page and Brin figured that with the proper algorithm a computer could rank the importance of Web pages and their relevance to any particular query. The algorithm treated links like academic citations: a page that was referred to by many different sites must be pretty important, especially if the sites doing the linking had themselves been judged important. Likewise, a page that was frequently described with the word *baseball* most likely had something to do with the sport. The algorithm for ranking Web pages' relevance was given the name PageRank in a nod to co-creator, Larry Page.

PageRank didn't just produce results far superior to other search engines', the two wrote in an academic paper about their project, but it produced results that agreed with "people's subjective idea of importance."[9] That is, the algorithm's recommendations appeared to jibe with what you yourself would have come up with if you were a search engine. This was true even though Google was giving generically useful answers. The search engine didn't necessarily know you from Adam, but imagine if it did? The more Google knew about the person searching—where she lived, what she liked to do—the better its search engine would perform when trying to answer user queries. No matter how intelligent an algorithm is, if it doesn't know whether you live near Berlin, Germany, or Berlin, Ohio, it's not likely to give the most useful answer to the search "bookstore in Berlin."

Google would become aggressive in compiling digital dossiers about its users to keep perfecting search, remembering each query and which results were deemed relevant. But even before such tracking became regular practice, Page and Brin used John McCarthy's sprawling Stanford Web site to run an experiment on personalization. McCarthy was freely sharing his papers and reviews, and links to other Web sites, as well as his short, biting opinions about current events. Using all this material, the two created a filter personalized to McCarthy; not surprisingly, when PageRank applied that filter to search results they were likely to be much more relevant to McCarthy than those produced by the generic algorithm. Personalized search engines, the two wrote in a paper coauthored with their academic advisors, "could save users a great deal of trouble by efficiently guessing a large part of their interests given simple input such as their bookmarks or home page." For example, "while there are many people on the Web named Mitchell, the No. 1 result is the home page of a colleague of John McCarthy named John Mitchell."[10] McCarthy had collected an unlikely honor: he was perhaps the first of a billion or more people to be profiled by Google so that it could generate personalized search results.

Page and Brin indeed are direct descendants of McCarthy and his artificial-intelligence-inflected computer science, although when the two were getting started in the mid-1990s, the field's ambitions had been severely clipped and McCarthy was known as a legend rather than a cutting-edge researcher. Like David Shaw's gifted investing computers, PageRank was a highly practical use of artificial intelligence techniques based on faster calculating and access to more data rather than a breakthrough in independent thinking. The computers running PageRank appeared to think like a person by working efficiently and incessantly as only a silicon-based machine can. "We don't always produce what people want," Page said early on. "It's really difficult. To do that you have to be smart—you have to understand everything in the world. In computer science, we call that artificial intelligence."[11]

Again, this application of AI may have fallen short of the high ambitions of those who believed they were on the verge of creating a new form of intelligence, but in terms of how we live the impact of the uncanny Google search engine was immense. The forty years between 1956, when McCarthy came up with the term *artificial intelligence*, and 1996, when Google arrived on the scene, encompassed a digital revolution. The pioneers of AI began their work on huge, expensive, memory-deprived mainframe computers. Only a select group of experts in business, academia, and government were capable of programming these colossi, and only a minuscule group of MIT undergraduates— the hackers—particularly wanted to. The general public experienced computers from afar, and were quite suspicious. IBM was well aware of this suspicion, McCarthy recalled with regret, and didn't conduct any AI research from 1959 to 1983. Their slogan attempted to lower the stakes: "data processing, not computing."[12]

By the time Google was invented, however, computers had become epically faster and cheaper; memory was becoming manageable as well. The PC had inserted itself into daily lives. Under those conditions—let alone today's world of pocket-size devices—AI research oozed with real-world implications. Anything that made computers act more "human" was of immediate interest either to regular folks in their daily lives or the elites in finance, as well as the military and government, which had highly specialized tasks for computers to perform. There was abundant enthusiasm for Shaw's computer algorithms, which discovered and exploited minor inefficiencies in the stock market to make billions of dollars, as well as for PageRank, which for the first time allowed hundreds of millions of Web users to go exactly where they wanted online.

When Page and Brin started asking hard questions about search, seemingly everything online was taking a turn toward the crass commercialism best exemplified by the blinking banner ads that had multiplied across the Web. In 1993, barely 1 percent of servers belonged to a .com domain, meaning commercial; by 1997, that share was 60 percent.[13] Web search was no different, filling up with opportunists looking to make a quick buck. Many of the leading companies preferred to manipulate results on behalf of advertisers rather than improve the underlying technology. In their paper introducing the ideas behind the Google search engine, Brin and Page express dismay that these companies wouldn't share their raw data so that the two researchers could test their ideas. Not surprisingly, the companies that had the information considered it too valuable to part with and weren't necessarily interested in what a couple of up-and-coming researchers might have to contribute.[14] Google would be different, they promised, with its data organized so that "researchers can come in quickly, process large chunks of the web, and produce interesting results that would have been very difficult to produce otherwise."

Entrepreneurs may have been content to treat search like just another service ripe for pillaging, but Page and Brin saw the stakes as so much higher. A search engine, more than other Web tools, only worked if it had a user's trust, which required transparency. Too much of what a search engine did occurred out of sight. If it steered you away from a certain site to please an advertiser, how, precisely, would you know? Even computer experts were hard-pressed to determine if the commercial search engine they were using was playing fair when it published its results. Again, Google would be different, a project run by academics who were committed to improving search for its own sake, not to make a quick buck. There would be no ads. In a paper the two delivered to a conference in Australia in 1998, Brin and Page took pains to explain why advertising inevitably corrupted search technology and had to be kept out. They offered both ethical and practical arguments.

The ethical obligation to run a Web search engine without advertising reflected an academic's belief in the importance of public access to information, which could be a matter of life and death. In a remarkable appendix to that 1998 paper, "The Anatomy of a Large-Scale Hypertextual Web Search Engine," Brin and Page offered an example of the danger to the public from search results that were tainted by advertising. As a test, they typed in the query "cellular phone" at all the prominent search sites. Only the Google prototype, they reported, returned a top result that was critical of cell phones, specifically a cautionary study about speaking on the phone while driving. PageRank didn't return the link in order to do the right thing, the two explained; it was simply conveying to its users what the Web thought were the most relevant links to someone interested in cell phones. The better question to ask was, Why didn't the other sites link to that study? Page and Brin's answer: "It is clear that a search engine which was taking money for showing cellular phone ads would have difficulty justifying the page that our system returned to its paying advertisers."[15]

Practically speaking, as well, advertising led to an inferior product. Brin and Page didn't point to obvious examples of manipulation like a search engine that promoted an advertiser's business undeservedly to the top of results, or, as in the cell phone example, buried results that might offend. Instead, they were struck by the opposite effect: that a search engine might tamper with useful results to punish an advertiser. There was the case of one prominent search site that appeared to intentionally ignore an airline in the results of a search of the airline's name. The airline had paid to have a prominent advertisement linked to such a query, and the owner of the search site must have feared that if the airline discovered that its name already appeared in front of users who were looking for it, then it might not bother to advertise at all. This meant, incredibly, that under certain conditions, more relevant search results were actually bad for business. What could be more offensive to a pair of artificial intelligence researchers than a system with a feedback loop that made things worse?

Brin and Page had put their finger on the Catch-22 of the search business: if results improved so much that they became uncannily accurate and precise—like a chess computer arriving at the single best move for a certain position—then advertising will have lost much of its purpose. You would be shown where to go based on the consensus "best result," and thus should have little interest in hearing what an advertiser wanted to tell you. OK, there might be a few ads to introduce a new product, or to try to persuade you to switch between brands, but this wasn't the basis of growing business. A search engine needed to sell something valuable—like reaching customers in a way competitors couldn't—if it wanted to make a lot of money. The bad incentives were clear: search companies would stop trying to improve their services for business reasons, which is why Page and Brin toward the end of their paper made the following assertion: "We believe the issue of advertising causes enough mixed incentives that it is crucial to have a competitive search engine that is transparent and in the academic realm."[16]

The Google search engine from the start was a monumental improvement on the other search engines and the good news spread across the Web purely by word of mouth, just as was the case with the Mosaic browser a few years earlier. Traffic to Google grew 50 percent a month, month after month.[17] Here was another Internet-defining creation from students barely in their twenties, who were working out of a university computer lab and reaching a worldwide audience through the university's wires and cables. Google was operating with a nest of different computers housed in cooling cabinets made out of Legos and through an Internet connection at the Bill Gates Computer Science building, a $38 million complex that opened in 1996 spurred by a donation of $6 million from Gates.[18] When Google's Internet connection appeared to reach its breaking point, a member of the team suggested switching a toggle so its computers could tap into the T3 line serving the entire university, which relieved the pressure for a time.[19]

A Web phenomenon like Google didn't remain a secret for long to the university (with that toggle flip, the search engine grew to use as much as half of Stanford's total bandwidth) or the news media or potential investors. What would the young computer scientists say to those who wanted to commercialize their magnificent new Web tool? Histories of Google tend to rush past what was an epic decision to abandon their conviction that advertising corrupted Web search. One account of these events put it this way: "Soon, the temptation to spin it off as a business was too great for the twenty-something cofounders to bear."[20] In his insider account of Google's history, *In the Plex*, the writer Steven Levy doesn't even credit the pull of temptation. "Page and Brin had launched their project as a stepping-stone to possible dissertations, but it was inevitable that they began to eye their creation as something that could make them money."[21] Inevitable.

Brin and Page have come to terms with their change in priorities by concluding that ordinary rules don't apply to brilliant engineers like themselves: Google would somehow remain a research project with noble aims even as it transformed itself into a business with a duty to make money. One explanation for how they managed this particular form of cognitive dissonance can be found in Page's obsession with Nikola Tesla, the brilliant Serbian-American inventor who died in relative obscurity. Tesla's superior ideas in electricity and other fields lost out to Thomas Edison and others, who had a superior business sense. Tesla's life became a cautionary tale. "I feel like he could've accomplished much more had he had more resources," Page said. "I think that was a good lesson. I didn't want to just invent things, I also wanted to make the world better, and in order to do that, you need to do more than just invent things."[22] A great scientist needed to tend to business, if only to be sure his ideas had a chance to take hold. Placating your investors—in Google's case by accepting advertising tied to search results—wasn't ethical compromise but a way to make the world a better place.

There is another way of explaining why Brin and Page reversed course. They were students at Stanford, after all, and at Stanford, you don't look for investors, investors look for you. Marc Andreessen cited his own experience at the University of Illinois working on the Mosaic browser to describe what was so different about Stanford. "Had we been at Stanford," he said, "of course we would have started a company. In the Midwest, no way. Not a chance."[23] As discussed earlier, Only when Andreessen flew to Silicon Valley did his entrepreneurial horizons broaden. There, under the guidance of a former Stanford engineering professor, Jim Clark, was he educated in the ways of investment and entrepreneurism. Together, they built a browser business, Netscape, which got the dot-com era started with a bang.

Stanford meant something very different to Brin and Page, who first met there in 1995, a year after Andreessen arrived. After impressive academic careers at the state universities of Maryland and Michigan, Brin and Page were joining the academically prestigious Stanford computer science department to complete PhDs. They had committed themselves to an academic life. They weren't blind, however, to the entrepreneurial legacy of Frederick Terman, what Brin called Stanford's "big history of company building."[24] Even before they became fast friends and research partners, there was an example of a wildly successful Web start-up on the Stanford campus, Yahoo, which was built by a pair of electrical engineering graduate students, Jerry Yang and David Filo, working in their spare time.

Yang and Filo created their first Web index in January 1994, which quickly became "Jerry and David's Guide to the World Wide Web" and then Yahoo, to keep track of the technical papers online related to their PhD research. Other electrical engineers found the index helpful and sent suggestions of cool things on the Web; the mission and the audience kept expanding. "All of a sudden both of them went from doing their graduate work to adding Web sites to their list for eight hours a day," Tim Brady, Yang's college roommate at Stanford, recalled. "As chance would have it, their thesis advisor was on sabbatical, so there was really no one looking after them, so it all worked. Had their advisor been there, it might not have happened."[25]

On the face of it, Yahoo, which incorporated in 1995, was a very different project from the expensive engineering projects that Terman had encouraged Stanford graduate students and faculty members to pursue at first within the university lab and later in the private sector. Those projects typically earned government research grants, which included overhead payments to Stanford, before being brought to market in partnership with large multinational corporations. In the case of Yahoo, Stanford graduate students were working on a marketable project that had nothing to do what they were researching—actually took time away from what they were researching. This development path raised an interesting question: would Stanford claim to own the index, since it was created on university equipment, even if it wasn't related to university research?

Brady, a recent Harvard Business School graduate and the first outside employee at Yahoo, recalled being nervous about this central, even existential, question for the company. But Yang and Filo reassured him. "Stanford is very progressive in that," Brady said. "Yahoo is far from the first startup that originated there and will be far from the last one. It was new enough, and it wasn't a specific technology; it was a brand."[26] Unlike the University of Illinois, which didn't bend to the will of its students, Stanford as designed by Terman was content to bet on its students, even when they were playing academic hooky. Brady explained the Stanford philosophy, as it applied to Yahoo. "They were smart enough to know that anything they would do to stifle it would kill it," he said of the university's administrators, "so their best hope was to just let it go and hope that Jerry and Dave gave money back later, which they did. They optimized their outcome, trust me."[27]

Indeed, after getting the go-ahead from the university, as Yang and Filo had predicted, Yahoo grew through an early investment by venture capitalists on Sand Hill Road. In April 1996, when Yahoo had its IPO, it reached a valuation of more than $700 million.[28] The next year, Yang and Filo each donated $1 million to Stanford to endow a professorship at the engineering school—at ages twenty-eight and thirty, they were the youngest donors to make such a significant donation since the school kept records.[29] And today, Yang is on the Stanford board of trustees; he and his wife, Akiko Yamazaki, have pledged $75 million to the university to expand interdisciplinary studies.[30] All was going according to Terman's plan. Stanford was encouraging its entrepreneurially minded students to start a business—playing the role, one might say, of the canny initial investor, providing crucial start-up equipment, overhead, and technical assistance. Instead of shares in Yahoo, the university produced goodwill that translated into donations, almost immediately. Stanford's reputation was soaring, right along with its endowment.

Brin and Page were a different case, however. Their idea for how to navigate the Web wasn't meant to assist their PhD research—it was their PhD research. They hadn't stumbled on a business idea, they were methodically wrestling with difficult questions involving data storage, data retrieval, and artificial intelligence. Their initial discomfort in starting a business from their research reflected how seriously they took these profound questions. There had been a countervailing belief system within the Stanford computer science department, represented by John McCarthy, among others, that considered commercial success inappropriate for a serious academic. A deep thinker, a true scientist, should stay at the lab and try to change the world from there. Only the short-term students—dropouts, even—were focused on making a fortune from the ideas they had picked up at school.

Google, like Yahoo, required substantial Internet bandwidth, especially as it became more popular, but the PageRank algorithm was PhD-level research being conducted at a world-class university. It could hardly have been the most ambitious or costly research project operating there; consider the massive particle accelerator that Terman had helped land for the university in the early 1960s. Couldn't support for Google be just another case of the university backing its scientists wherever their research took them, money be damned? Wasn't studying the burgeoning World Wide Web a subject worthy of academic research and university investment?

From that perspective, Stanford's backing of Google shouldn't be seen as acting as a venture capitalist but rather as a nonprofit institution committed to supporting cutting-edge research. In fact, what turned Brin and Page into entrepreneurs wasn't Stanford's early "investment" in their research project, but rather its decision to *stop* supporting the fast-growing Google project. The withdrawal of support in the summer of 1998 meant that Brin and Page would have to fend for themselves. "The resources just weren't there in the academic environment," Brin recalled. "We decided, 'Hey, there's a lot of commercial interest here. People will give us a lot of money to solve this problem of search. Why don't we go and do it commercially?'"[31] Hector Garcia-Molina,

Brin's advisor at Stanford, has a slightly different view of the events of this period, seeing the Google founders as being pulled toward the market rather than being pushed there by Stanford. "It wasn't like our lights were dimming when they would run the crawler," he said of Google's supposed overreliance on the university's equipment. "I think it would have made a great thesis. I think their families were behind them to get PhDs, too. But doing a company became too much of an attraction."[32]

In August 1998, when Brin and Page were exploring how to keep Google going, they sought the advice of David Cheriton, a professor wise in the ways of start-ups, who worked a few doors down on the fourth floor of Margaret Jacks Hall. Cheriton had made a small fortune a couple of years earlier when Cisco paid $220 million for a company he created with Andy Bechtolsheim, a former Stanford computer science graduate student who had dropped out to help start Sun Microsystems. Cheriton encouraged Brin and Page to seek funding for Google, looping in Bechtolsheim. "If you have a baby, you need to raise it," the professor told the students.[33] The next morning, Brin and Page took a meeting with Bechtolsheim at Cheriton's house.

Brin and Page demonstrated Google on a laptop, and, as Brin recalled, Bechtolsheim interrupted to say, "Oh, we could go on talking, but why don't I just write you a check?" He ran to his Porsche, cut a $100,000 check made out to Google Inc., and handed it to Brin and Page. They explained to him that there was no Google Inc., and therefore no account in which to deposit the funds. No matter, Bechtolsheim replied, "Well, when you do, stick it in there."[34] No terms were discussed. It was unclear what exactly he was buying with his money, or if it was only a loan. But just like that, Bechtolsheim had turned two graduate students' curiosity about starting a business into something tangible and real. "It was like, wow, maybe we really should start a company now," Brin said. "The check sat in my desktop drawer for a month. I was afraid I'd lose it. But until it really happened, until then, it had sort of been this intermediate state. Things hadn't really happened yet. But when he wrote the check—well, it certainly does speed things up."[35]

Google incorporated on September 7, 1998, and Cheriton would become one of four early investors in the company, each said to have written checks for $250,000. In addition to Bechtolsheim, who supplemented his initial investment, and Cheriton, there was Ram Shriram, an angel investor who was introduced to Brin and Page through a different Stanford computer science professor, Jeffrey Ullman, who also encouraged the two graduate students to go into business. You can always come back and complete your degree, Ullman assured them. Shriram signed up the fourth investor, Jeff Bezos.[36] The seduction of Brin and Page wasn't simply about greed, or having the resources necessary to "raise their baby" and thus avoid becoming modern-day Teslas, whose best ideas never reached maturity. There also was the thrill for these star students of getting A's in real life, not just in the classroom. Cheriton, Bechtolsheim, Shriram, and Bezos were talented, wealthy men who were clearly impressed by the business potential of PageRank. The multimillion-dollar investment round that Brin and Page pursued next would be another big fat gold star right on top of their business plan, one that everyone could see.

While making the rounds to pitch the Google search engine to eager Sand Hill Road firms nearby, Brin and Page were impressed both by John Doerr of Kleiner Perkins, which had invested in Netscape and Amazon, and Mike Moritz of Sequoia Capital, which had backed Yahoo. What to do? Why not work with both? Bechtolsheim told them there was a "zero percent possibility" that this would happen: hypercompetitive, hypercapitalist investors, he explained, don't like to share their quarry.[37] Those words were all the motivation Page and Brin needed. When the road show was over, the two firms agreed to split a $25 million investment at a valuation of $100 million. Making the sale wasn't exactly easy, Page said, but he and Brin had the advantage of "being at Stanford and having a product that was really good. . . . It was much better than anything else, and so we would show them and say, 'Hey we've got this great thing. It works better than anything out there. What do we do?'"[38]

As these events illustrate, Google's founders displayed a particular mix of iconoclasm and respect for status. They clearly didn't care for the rules for how VC funding was done, but that rebellion was in pursuit of being the rare start-up that could win over two competing firms. Page and Brin set an idealistic goal for the company—"To organize the world's information, making it universally accessible and useful"—and because of the difficulty of that task, they insisted on hiring applicants from a narrow range of top-tier universities who met the rigorous academic standards common to Silicon Valley. Google intended to measure the "pure" intelligence of the candidates for its jobs and liked to deploy riddles on the spot as a way of detecting genius.

It's hard to know exactly where the attraction lay for the Termans or David Shaw or Jeff Bezos or the Google founders in trying to measure intelligence so exactly. That is to say, did they want to assign numbers to intelligence because they liked the efficiency of using an algorithm to make judgments about hiring and promotion, or did they sincerely think intelligence was best rendered as a number, the way a computer's processor could be assigned a speed? Whatever the motivation, they were all sticklers about getting as much relevant data as possible about a job applicant's brain power and acting on that information. Thus Terman would routinely step in to reject a candidate for a professorship in a field like biology based on his undergraduate grades in calculus. Similarly, Page and Brin would insist on reviewing the standardized test scores of each Google applicant. A candidate with low test scores could be instantly disqualified. As recently as 2010, Page had the ability to examine the dossier of grades and test scores for every Google applicant. Occasionally, he did a spot check on a candidate's credentials, "to ask what is the real quality of person we're hiring." Google wallowed in grades and scores, whether ancient or newly manufactured. For example, each answer during the hiring interview was rated on 4.0 scale—just like college—and scoring an average below 3.0 meant you didn't pass.

Again, the values of the computer lab were imported into a company. "We just hired people like us," Page explained. There were classes, cafeterias, esteemed visitors like McCarthy, who showed up early on to talk to employees. Sometimes programmers would take the final leap and sleep at work. "The fact of the matter is that for some people living here makes sense," says Eric Schmidt, the older computer science PhD who was brought in to run Google in 2001. "Their friends are here, it's what they're familiar with, and the things they do here are very similar to what they did in college."[39]

In the year before the dot-com crash, there was a broad acceptance of and patience for the strategy of putting together the best computer lab, giving it a goal, and then figuring out how to turn a profit. The $25 million investment in Google Inc. came in June 1999, and at the sparsely attended press conference in the Gates computer building on the Stanford campus, Brin was asked how Google intended to make money. He replied, "Our goal is to maximize the search experience, not to maximize the revenues from search."[40] The next month Google made a grudging acknowledgment of those newly arrived investors by assigning its first staffer to study how to bring advertising to the site. Generally speaking, however, the company's brilliant founders felt free to focus on making the search engine better and adding more users— as if nothing had changed from their university days. When the subject of advertising came up, Brin and Page insisted that it wouldn't be allowed to threaten the integrity of PageRank, which was sacrosanct. Licensing the company's search technology seemed like the better business. The purer business. Advertising remained an ethically suspect backwater.

In late 2000, however, after the dot-com bubble burst, Google was under real pressure to perform—the company still hadn't registered a profit despite 70 million daily visitors. Cheriton, among others, made his disappointment known, joking that all he would have to show for his investment was Google swag, including what he called "the world's most expensive T-shirt."[41] Some investors started to suggest that they might withdraw their money if others could be found to step in—the specter of failure for the first time hung over Google's founders. Doerr and Moritz used this stressful time to again insist that Google have an outside chief executive, something Page and Brin had agreed to but pushed off at the start. This time, however, they had to accept, and after many interviews arrived at Schmidt, a forty-five-year-old engineer who had already led a large tech company.[42]

With investors suddenly skeptical of how Google was operating as a business, advertising became a higher priority and Page and Brin assigned a number of talented executives to come up with an answer. After a series of small failures, Google in 2002 developed AdWords Select, an improved system for delivering ads next to search results that seemed to have solved a multidimensional puzzle. The ads ran well to the side so no one would confuse ads for search results, a taboo for Brin and Page. Under Page's prodding, AdWords Select embraced an automated sales procedure that could scale easily, stepping in for the company's human sales team. Anyone with a credit card could use the system and advertisers were only charged by Google for a click-through, that is, when someone left the search results page for their landing page, as opposed to being charged based on how many people were shown the ad. The system tapped into all Google knew about Web pages and its users to guide a customer to the strategy that would work best. The auction system for deciding which ads would appear next to certain searches relied on feedback loops that produced better outcomes for everyone involved. Winning bidders were refunded the difference between their bids and the runner-up and, by treating bidders fairly in this way, Google gave them the confidence to be bold with their bids, spending as much as they could afford. Likewise, Google ranked ads in part based on how effective they were, not simply rewarding the highest bid. This encouraged advertisers to make ads more useful, thus pleasing Google's users.

In each case, the ethical system proved the most effective. Google's first profitable year was 2001: $10 million. The next year, when AdWords Select arrived, profits were $185 million. Revenue was growing fast, too. The next year, Google introduced another advertising system, AdSense, which allowed small-scale Web sites to carry AdWords-style ads based on the kind of material they published, guided by the data Google had collected about who clicked on which ads. Google took roughly a third of what the advertising brought in, with the rest going to Web site publishers, many of whom otherwise never would have had the wherewithal to run ads.[43] In 2004, when Google went public and had to show its books, investors could see exactly how profitable the company was, even in what clearly were lean years for dot-com companies. With the company on firmer footing, Brin and Page's iconoclasm returned during the IPO, which was designed to work around investment banks by selling shares directly to individuals through an auction. (The original filing sought to raise $2,718,281,828, an allusion to the mathematic constant e, 2.718281828. . . .)[44]

The two founders were confident they had threaded the needle, finding a way to bring in lots of revenue with ads that were distinct from the search engine proper, thus avoiding the harm they had so carefully documented in other commercial search engines that accepted advertising. The ads displayed next to Google search results were so well targeted that users didn't even know the ads were there unless they were looking for them, Brin said. "Do you know the most common feedback, honestly?" he asked. "It's 'What ads'? People either haven't done searches that bring them up or haven't noticed them. Or the third possibility is that they brought up the ads and they did notice them and they forgot about them, which I think is the most likely scenario." Google ran an A/B test to learn more, displaying an ad-free experience for some users, while others saw ads. The visitors who were shown ads, the test revealed, used Google more than those in the ad-free group, leading to the happy conclusion that people liked Google's ads.

However, Brin and Page's critique of advertising relied on some clear guideposts for judging if search was being harmed by advertising and the search for profits and whether someone who stayed on the site longer wasn't one of them. One issue to investigate, according to the two when they were in graduate school, was whether advertising stifled innovation. In terms of queries about people, places, and things (that have nothing to do with commerce), Google has undoubtedly been improving, approaching the kind of uncanny results that the founders originally envisioned. Google effectively anticipates what you are typing and usually can provide a précis on any topic of your choosing, relying on Wikipedia and other deep online resources as well as the information the search engine has collected itself.

In today's version of Google, however, ads are no longer automatically segregated from results; they frequently appear directly above them. Enter an open-ended search term like "bicycles" and Google will direct you to some neutral sites that provide some perspective on the subject, but there may also be ads, and above those ads may be a row of shopping opportunities. The effect can feel highly commercialized. Here is the explanation Google offers for the sponsored shopping deals that frequently appear: "Based on your search query, we think you are trying to find a product. Clicking in this box will show you results from providers who can fulfill your request. Google may be compensated by some of these providers." There is a bit of a chicken-egg aspect to that explanation. Is Google merely reflecting the Web by focusing on commerce and assuming that someone who makes a generic search must be "trying to find a product"? Or is it helping to shape this reality through its decisions about how to organize information? A search engine was supposed to be an honest, disinterested broker between a seeker and what is available on the Web, according to Brin and Page. Google obviously isn't that anymore.

Another point that Brin and Page warned against as graduate-student foes of commercial search was the lack of transparency. Research at a university is typically published so that it can be reproduced and scrutinized and improved, while research done for a business is done in secret. The latter leads to what the two back in 1998 called "a black art," which can be used for good or ill depending on who knows the magic formula.[45] Today, Google Inc. jealously guards its secrets, PageRank first among them, and has offered a couple of justifications. To start, there are competitive reasons. Schmidt had battled with Microsoft in his previous job at Novell, and thus advised Page and Brin to keep as much secret as possible to avoid giving a competitor any angle to exploit. Netscape had experienced how Microsoft could quickly take over a market—like browsers—by applying its tremendous resources to reverse engineer a problem. Even revenue breakdowns, by Schmidt's lights, could offer clues as to how Google works. There were the practical arguments, too, for practicing secrecy as it related to search. If the public knew even in broad strokes how PageRank made its calculations some enterprising souls would be able to game it and search results would suffer. Without giving opportunists that advantage, Google still battles against companies and individuals who practice what is called "search engine optimization."

Page and Brin have done a true 180 on the topic—transparency was once the path to the best results, and now they view it as an obstacle. Moving from a research lab to a start-up obviously has changed their perspective. If Google still operated transparently and noncommercially in a university setting, then when some companies tried to trick the search engine the consequences wouldn't be so dire. An open system would be self-correcting: anyone keeping up with the project could spot the dirty tactics and propose a solution. Wikipedia, for example, has created a system that relies on human volunteers as well as "bots" to control vandalism or self-motivated editing. It tolerates some slip-ups but generally has succeeded in sapping the will of troublemakers.

Such deception is much more threatening to a commercial entity like

Google, however. It goes right to the heart of the company's pitch to advertisers and Web site owners alike that it runs a closed system that is beyond significant manipulation. Beyond outside comprehension, really. The algorithm does its job secretly and fairly—that is Google's promise. Its results are above reproach and the only way to deviate from this preordained "correct" outcome—to make an extraordinary impact for your business in search results—is to pay for a reasonably priced, demonstrably effective ad. This is good for search, which is in the trusted hands of the experts who work for Google. And good for Google.

In this way, Google tries to cultivate in the public the same cognitive dissonance it displays about whether it is a high-minded research lab or a profit-minded business. The entity that collects all that information, Google assures you, isn't some company spying on your behavior to sell ads, but a research lab trying to improve your search experience and only show you ads that you'll want to see. The same goes for the broad definition of "fair use" that Google favors, which would permit it to collect all manner of copyrighted material—movies, music, books—in its computer storage to be included in its search results. This is not because it wants to profit from others' work, but because of its high-minded purpose of organizing the world's information and making it accessible. Same goes for its opposition to the European "right to be forgotten," which requires that someone mentioned in an old news article be able to demand that that article be excluded from search results.

There was a final experiment that Brin and Page proposed twenty years earlier to gauge the influence of advertising on search: type in the query "cellular phone," they suggested, and see what results appear. Running that search on Google (today *cell phone* is the preferred term) produces what the two predicted would be the case for any search engine dependent on advertising. You do not find a negative Web page anywhere near the top of the results. First, there is a banner of links to buy cell phones, with the label "sponsored." Then there are two ads, clearly labeled as such, and then a map showing you a few places to buy phones. Then the proper listings begin—Wikipedia's article, Walmart's article. Then cell phones in the news. Then, about a dozen or so items down is a *Scientific American* article from 2008,

"Mind Control by Cell Phone," that is mildly cautionary. Brin and Page's paper didn't claim that the absence of critical links about cell phones directly proved corruption in the search process. They offered a rational explanation: "A search engine which was taking money for showing cellular phone ads would have difficulty justifying the page that our system returned to its paying advertisers." This subtle point, coming from would-be academics, recognized that you can produce bad outcomes without having that intention. You may just be doing what it takes to make your business run and "being evil" in the process, without even noticing.

All of which raises some intriguing alternative history: Would Brin and Page have remained true to noncommercial search had they met at a top-class school with a less go-go culture, like, say, MIT? Could they and we have avoided this reimagining of Google's relationship with its users? Or would we instead be talking about a different pair of brilliant Stanford graduate students who researched how to improve Web search and inevitably found their way to Sand Hill Road, while our alternative-universe Brin and Page would be obscure but distinguished computer science professors?

Looking back two decades, Google's rise from research project to Google Inc. may represent the last gasp of the Terman model at Stanford. Certainly innumerable start-ups have their roots in Stanford, including Instagram, but the university's resources aren't nearly as significant. Less than a decade from Google's founding, an ambitious hacker could make his mark from his dorm room, using just the money in his pocket. Mark Zuckerberg was a twenty-one-year-old chief executive of Facebook when he returned to his old Harvard computer science class to tell the students how lucky he was to have started his company in the era he did. "The fact that we could sort of rent machines for, you know, like $100 a month and use that to scale up to a point where we had 300,000 users is, is pretty cool and it's a pretty unique thing . . . that's going on in technology right now." A decade earlier, he said, eBay had to run off of two $25,000 machines from the start. Google was somewhere in the middle of these two poles and had some serious bills to pay.[46]

If Google was a last gasp for Stanford, what a last gasp it was.

Because PageRank was developed at Stanford, the technology behind Google is owned by Stanford, which exclusively licensed it to Brin and Page for a portion of Google stock, the last of which Stanford sold in 2005 for a total payout of $336 million.[47] The Google-Stanford connections are many and varied, as Terman imagined—students, faculty, alumni, and the university all helping each other make a fortune. Google has returned those good turns from Stanford in so many ways, from $1 million a year directly donated to the computer science department, to money donated to honor the memory of an inspiring computer science professor, to hiring Stanford graduates, and on and on. The recently departed president of Stanford, John Hennessy, is on Google's board.

But, just as the building named after Terman was torn down to make way for an entrepreneur donor, this Stanford model for success ended up being replaced by other systems offering financial and social support, whether that meant informal networks of already successful entrepreneurs or specialized tech incubators like Y Combinator and Tech Stars. One Stanford graduate, Peter Thiel, proposed cutting through the pretense. Why pay tuition if you weren't there to study? "College can be good for learning about what's been done before, but it can also discourage you from doing something new," his organization, the Thiel Fellowship, explains on its Web site. "The hardest thing about being a young entrepreneur is that you haven't met everyone you'll need to know to make your venture succeed. We can help connect you—to investors, partners, prospective customers—in Silicon Valley and beyond."[48]

This was Thielism in its raw form—no half measures or false solicitousness. Stanford already was breaking down the purpose of academia by being so eager to turn research into profits, which it would share indirectly or directly. Why keep up the charade about university life? Why not go all the way and accept that market success was all that really mattered?

7. PETER THIEL

"Monopolists lie to protect themselves"

I f moguls like Gates, Andreessen, and Zuckerberg started as ambitious young hackers who discovered that their obsession with computers could lead to great wealth and power, Peter Thiel was something like the opposite—an ambitious young man who discovered that his obsession with great wealth and power would lead to computers. After graduating Stanford with both a B.A. in philosophy and a law degree in the early 1990s, Thiel pursued a conventional career in finance, briefly working for a blue-chip corporate law firm and then shifting to become a Wall Street banker. By 1998, however, Thiel was running his own small investment firm back in the Bay Area, where the dot-com gold rush was in full effect.[1] He had already helped friends with start-ups that had "sort of blown up in catastrophic ways" and was determined that, the next time, he would be involved from the beginning and do things differently.[2] Later that year, Thiel returned to the Stanford campus to give a guest lecture—fittingly, his talk was held in a small room in the not-yet-torn-down Frederick Terman Engineering Center. It was there he would find the ambitious hacker to help him on his way.[3]

The subject of Thiel's talk was currency trading, his investment specialty, but inevitably he touched on a broader political point—

how market globalization would lead to political freedom. "It was a topic dear to his heart. . . . Peter's philosophical underpinnings were influenced by accounts of totalitarian oppression such as the works of Aleksandr Solzhenitsyn," writes Eric Jackson, who had just graduated from Stanford and knew Thiel as the founder of the conservative campus magazine, *Stanford Review*, where he had been on the staff.[4] Among the small handful in the audience was Max Levchin, an intense, libertarian-leaning Soviet Jewish émigré who had recently graduated with a computer science degree from the University of Illinois at Urbana-Champaign.

Levchin had arrived on the Illinois campus in 1993, just as Marc Andreessen was leaving, declaring the Midwest an entrepreneur-free zone and taking the best programmers with him to Silicon Valley. Levchin was among the undergraduates hired as their replacements at the supercomputing lab where the Mosaic Web browser was developed, but he wouldn't be as naive about business opportunities. He started companies while still in school, and was clamoring to head to Palo Alto as soon as he could convince his immigrant parents that he wouldn't be pursuing a higher degree. Levchin's life goals had been fundamentally transformed by Andreessen's example from "somebody who thought of myself first and foremost as a scientist/future academic to someone who thought there's no better way to be than to create businesses. Be an entrepreneur, that's what it's all about."[5]

Upon graduation, Levchin was invited by a friend from the computer lab to stay on the floor in his small Palo Alto apartment, which lacked air-conditioning. To say cool in the hot summer of 1998, Levchin snuck into talks on the Stanford campus, which is how he says he found himself listening to Thiel discuss financial markets and political freedom. "I went to see this random lecture at Stanford—given by a guy named Peter, who I had heard about but never met before," he recalls. Because there were so few people in the audience, Levchin was able to buttonhole Thiel afterward. What do you do? Thiel asked. I start companies, Levchin replied. That's great, I invest in companies, Thiel answered. The two committed to working together on a startup. "We hit it off really quickly—I have this IQ bias—anybody really smart I will figure out a way to deal with," Levchin says. "When we met, we sort of hung out socially, and then one night we had this showdown where we sat around in this café for like eight hours and traded puzzles to see who could solve puzzles faster—just this nonstop mental beating up on each other. I think after that we realized that we each couldn't be total idiots since we could solve puzzles pretty quickly."[6] It was a match made in heaven, or in Terman's Stanford, at least. Two young strivers—one proficient in the ways of computers, the other in the ways of law and finance—but each eager to promote radical change in a society they barely understood.

After some fumbling with other ideas, Thiel and Levchin came up with PayPal, which would create "a new Internet currency to replace the U.S. dollar." Thiel pitched the idea as fulfilling his anti-government dream of a global market that protects the welfare of the public by empowering them as consumers: "What we're calling 'convenient' for American users will be revolutionary for the developing world. Many of these countries' governments play fast and loose with their currencies. . . . Most of the ordinary people there never have an opportunity to open an offshore account or to get their hands on more than a few bills of a stable currency like U.S. dollars."[7] But there was that other thing, too, the chance to make a fortune. If you happened to create and own a new digital currency, you could collect a cut from each online transaction. You would become the middleman of e-commerce, as David Shaw and Jeff Bezos had sketched out, without incurring a retailer's burden of keeping track of orders, maintaining warehouses, and making deliveries.

There, in microcosm, were the two sides of Thiel, and two sides of disruptive Silicon Valley values: the self-proclaimed advocate for personal liberation who dreams of overthrowing the current order, as well as the ruthless entrepreneur dreaming of making a large fortune at Internet speed. Rather than those two identities' being in conflict, Thiel discovered, each can bolster the other. An appetite for the utopian social destruction represents a higher purpose than pure greed as your employees sacrifice to build your business. On the flip side, business success becomes vindication of all that destruction and proof that the right man is in charge.

Thiel was born in Frankfurt-am-Main in 1967, to German parents. His peripatetic father was an engineer who traveled around the world, briefly settling in, among other places, Cleveland, Ohio, where he studied at Case Western Reserve University and where Peter first arrived in the United States as a one-year-old and became an American citizen.[8] At age six, he spent two and a half years in apartheid South Africa and South West Africa (current-day Namibia), where his father worked for a uranium mining company. When Peter was ten, the family settled down in Foster City, outside San Francisco.[9] By Thiel's account, his was a conventional, if overachieving, upbringing: "My path was so tracked that in my eighth-grade yearbook, one of my friends predicted—accurately—that four years later I would enter Stanford as a sophomore."[10]

At Stanford, Thiel studied philosophy and began to separate even further from the pack. In June 1987, he started the right-wing publication *Stanford Review*, which Thiel reckons was his first entrepreneurial venture, and a successful one at that. He introduced the *Review* to give voice to young conservatives, who he felt were not being heard during the divisive debates of the day, like whether to allow military recruiters on the Stanford campus, or in the highly publicized fiasco over whether Stanford should agree to house the Ronald Reagan presidential library. Thiel put together the first issues of the *Review* in his dorm room with the help of a high school friend, but in barely a year, there were forty staffers who distributed twelve thousand copies of each issue.[11] Three decades later, the magazine still flourishes, with ample financial support from former editors, many of whom have reached the heights of Silicon Valley by following in Thiel's footsteps.

The topic that sustained the *Review* from the early days was the fight against "multiculturalism," the movement to make higher education more inclusive and welcoming to women, gays, blacks, and Latinos. Multiculturalism championed two main causes at Stanford in the late 1980s. The first was to overthrow a curriculum of "dead white males"—otherwise known as the Western canon—by insisting that Stanford's required courses on literature, philosophy, and art include more racial and gender diversity among the writers and artists being taught. The second was to demand that the university admit more women and minorities and protect them on campus by, among other things, aggressively punishing hate-filled student behavior.

The *Review* saw these efforts as trampling on university traditions, whether that be the tradition that students can say whatever they want on campus, including something hateful, or the merit-based system whereby the academy anoints the Great Writers and selects who can study at Stanford. When these two causes were taken together, it appeared to Thiel and his comrades, multiculturalism had the single, overarching purpose of coddling the weak. Stanford students, some of the most talented young people of their generation, shouldn't need to be shielded from occasional strong language or from the uncomfortable fact that the world's best thinkers were not neatly divided among various racial, sexual, ethnic, and social groups.

In *The Diversity Myth*, Thiel's look back at those times written with his friend David O. Sacks, a former editor of the *Review*, Thiel and Sacks describe a campus that has lost its grip on reality.[12] For example, they describe how women on campus didn't have it any harder than men, and mock a professor who argued that "women are so oppressed that we don't even know how oppressed we are—there are layers upon layers of institutionalized oppression." Not very intellectually rigorous, they write, since under this argument, "oppression was both pervasive and undetectable. Once again, sexism . . . had been defined as to be nonfalsifiable."[13] The same sleight of hand occurred on behalf of racial minorities. "The primary problem for multiculturalists is that there are almost no real racists at Stanford or, for that matter, in America's younger generation," the two wrote back in 1995. "The few exceptions, like the 'skinheads,' are highly visible (precisely because they are so few) and are not often spotted at elite schools like Stanford."[14] Ditto for gays and Latinos. All were being conned by their leaders to think that they were oppressed in order to avoid the burden of personal responsibility. There was class envy, too: "In the multicultural allegory, the wealthy are per se oppressive, because their success creates misfortunes for others. The other two possible causes of poverty—bad fortune or bad choices—are rejected a priori. Quite naturally, multiculturalists conveniently overlook the fact that without productive people paying taxes, there could be no welfare for the poor."[15]

Interestingly, the root conflicts of Stanford's culture wars—"merit" versus diversity (or lack thereof), and "freedom" versus tolerance (or lack thereof)—are precisely the ones that weigh so heavily on Silicon Valley today. Questions like: How do we bring more women and racial minorities into the tech world? Should we even try? And do we need limits on speech to make the Internet more welcoming to women and minorities who are frequently threatened online by hate-filled comments? Thiel and Sacks then, and now, resist official interventions on behalf of tolerance and diversity. Their vision at the time was of a campus of strong individuals who navigate their way in the world, judged by their deeds and their deeds alone, not asking for help nor wasting any time to take offense. That's pretty much the vision Silicon Valley is selling now, too.

An early incident involving graffiti on a picture of Beethoven became a cause for the *Review*. Two white students, Gus and Ben, had a heated discussion with a black student at Ujamaa House, Stanford's African American–themed dormitory, that culminated in the black student, B.J., claiming that Beethoven was African.[16] Thiel and Sacks pick up the story in *The Diversity Myth*: "The following evening, Ben noticed a Stanford Symphony recruiting poster featuring a picture of none other than Beethoven himself (in the poster, Beethoven appeared white). Inebriated, Gus and Ben used crayons to color in the Beethoven flier with the stereotypical features of a black man—brown face, curly black hair, enlarged lips—and posted the flyer on a 'food for thought' bulletin board adjacent to B.J.'s door."[17]

Thiel and Sacks agreed that the drawing was "certainly in bad taste," and that, undoubtedly, "the black residents did not consider the satire amusing," but they also detected hypocrisy in the reaction: "Overnight, with the most minute shift in inflection, the symbolic significance of the claim that Beethoven was black had changed 180 degrees—from a source of multicultural pride to a point of multicultural derision." The importance of context—who says what when—appeared to mystify Thiel and Sacks. "B.J.'s words and the two white students' drawing had said precisely the same thing. Nevertheless, the fact that these students were of different races made B.J.'s expression legitimate and the white students' something of a crime."[18] Is their description of these events another provocation? That is, are they "trolling" minority students, to use the popular term for online harassment, or did Thiel and Sacks sincerely think there was no difference in how a black student might say he believes that Beethoven was black, and how a drunk white student might? In a sense, the answer doesn't matter. The message is clear either way—we are tired of all this complaining from women and minorities and don't dare tell me what I can say.

The two freshmen who defaced the Beethoven poster were kicked out of student housing for the fall and winter quarters, but avoided any additional punishment when the university's top lawyer concluded that the racist graffiti was protected speech.[19] This solution left neither side happy: people sympathetic to the black student in the case were angry that the university ultimately concluded that painting stereotypical big lips, big nose, and an afro on a poster left next to a black student's room was permissible speech on the Stanford campus. People more sympathetic to the white students, including the *Review* editorial team, were angered that the stakes had become so high that a pair of freshmen were in danger of being expelled from Stanford over what they saw as at worst a joke in poor taste, which was made while intoxicated.

The university itself was left flailing, unable to find a workable middle path. During these tense events, tetchy Stanford administrators, as we've seen, removed a joke about a cheap Jew and a cheap Scotsman on an Internet Usenet group, which led John McCarthy to rally his computer science peers to fight for an uncensored Internet. Their victory in that fight, McCarthy says, laid the groundwork for the anything-goes Internet we live with today, which he considered one of his greatest accomplishments. Though the *Review* crowd didn't advocate for an unfettered Internet from a hacker's perspective, they shared the view that the Internet—and offline life, too—should be beyond the reach of meddling authorities. They, like the hackers, insisted they were living in a world where the effects of racism and sexism could be willed away, if they even existed in the first place.

There were other free-speech martyrs championed by Thiel and the *Review*, including Keith Rabois, a former editor at the *Review* who attended Stanford Law School with Thiel. Walking by a dorm that was led by a resident fellow the *Review* editors considered overly politically correct, Rabois yelled toward the home of the resident fellow there, "Faggot! Faggot! Hope you die of AIDS!" The hateful words were intended to educate the students who witnessed them about the extent of free speech protections on campus, Rabois explained.[20] "The entire point was to expose these freshman ears to very offensive speech," he wrote in a column for the *Stanford Daily* defending his actions. "Admittedly, the comments made were not very articulate, not very intellectual, nor profound. The intention was for the speech to be outrageous enough to provoke a thought of 'Wow, if he can say that, I guess I can say a little more than I thought.'"[21]

When administrators wanted to talk with Rabois about the incident, he told them he wasn't interested, confident he was beyond their discipline. "My time's too valuable," he told them. "Whatever was said was certainly legitimate criticism. I'm a first-year law student. I know exactly what you can say and what you can't."[22] Later, Rabois was hounded out of Stanford Law School by his peers, rather than school administrators, and transferred to the University of Chicago. "For all practical purposes, he was expelled," Thiel and Sacks write.[23] He reunited with Thiel at PayPal, one of at least ten former *Stanford Review* editors who "had played a vital role in shaping the direction of the company."[24]

There was a complicating factor about this particular campus incident, however: Rabois and Thiel were themselves gay, though both were still well in the closet. Homophobia was prevalent on college campuses throughout those years, and not just among conservative students, but especially among them. No doubt, the pressure of hiding who they were played out in bizarre tales like the "faggot, die" episode and in the *Review*'s criticism of an openly gay lecturer in computer science, Stuart Reges, who now teaches at the University of Washington. The *Review* took Reges to task for a comment he posted to a bulletin board at Stanford calling on gay students to be more vocal on campus to show administrators the hostility they faced. He proposed staging a kiss-in outside a gym known for homophobic incidents, or having gay couples attend fraternity parties to "find out just how open they are." This fit the pattern Thiel and Sacks observed among women and minority groups on campus—"the need to manufacture incidents indicates that there is not much of a problem."[25] How else could a libertarian-leaning publication like the *Review* oppose individual sexual freedom? Gay rights had to be seen as yet another minority group's attempt at special treatment.

Thiel was publicly outed as gay in 2007 by the Silicon Valley–based Web site ValleyWag, one of the Gawker family of Web sites that revel in making public the private lives of celebrities, or the sort of well known.[26] The headline was an exuberant "Peter Thiel is totally gay, people." Thiel has described the experience as like facing a terrorist attack: "I don't understand the psychology of people who would kill themselves and blow up buildings, and I don't understand people who would spend their lives being angry; it just seems unhealthy. . . . Terrorism is obviously a charged analogy, but it's like terrorism in that you're trying to be gratuitously meaner and more sensational than the next person, like a terrorist who is trying to stand out and shock people. It becomes this unhealthy dynamic where it just becomes about shocking people."[27]

Already a wealthy investor at that point, Thiel began a long-term strategy to turn the tables, helping to bankrupt Gawker Media by spending what he said was as much as $10 million to sponsor lawsuits against the online publisher. Nearly a decade after the ValleyWag item appeared, Thiel drew blood with a particularly potent grievance from the professional wrestler known as Hulk Hogan. Hogan, whose given name is Terry Bolea, sued Gawker over the publication of a tape showing him having sex with his friend's wife. When a Florida jury awarded Hogan $140 million in damages, Gawker had suffered a mortal blow.[28] Thiel says he didn't sponsor the lawsuits to exact revenge against an institution that did him wrong but, rather, to make the world a better place. His sub rosa campaign against Gawker, he said, was "one of my greater philanthropic things that I've done. I think of it in those terms."[29]

During the 2016 presidential campaign, Thiel was forced to confront some of the more extreme arguments in *The Diversity Myth*—like its description of date rape as sometimes masking "seductions that are later regretted" by the women making the accusations.[30] Thiel apologized for those comments, which he said have been taken, incorrectly, to mean he didn't think rape in all forms is a crime.[31] But he didn't renounce the book in its entirety, which is filled with small cruelties that don't lend themselves to headlines. As *The Diversity Myth* recounts in detail, the *Review* was always trying to get under the skin of its political opponents. If there weren't any wrenching disputes over race, gender, and class on the Stanford campus, the magazine would concoct them. There usually was some issue at stake—a professor's alleged abuse of power; conservative students' being denied a chance to get their views across—but take a step back and these provocations appear to be little more than an excuse to be mean.

There was the time that the *Review* identified an undergraduate class, Drama 113 (Group Communications), that seemed to lack the intellectual rigor befitting Stanford. Incredibly, to the *Review* editors, the instructor told her students at the start: "I am not interested in facts—I care about how you feel."[32] Sacks and other *Review* editors decided to infiltrate the class and start taking notes. One day in class, they reported, a Latina student became emotional while describing her working-class upbringing. "Trying to be helpful, an upper-middle-class person asked whether Tamara wanted her children to be members of the upper-middle class," Thiel and Sacks write. "Without a moment's hesitation, Tamara replied: 'absolutely not'; her children would grow up in a lower economic class as well. Several students responded: Then how could she be sorry for her own childhood?" They never did get an answer, Thiel and Sacks write matter-of-factly, "because the distraught young woman promptly ran out of class," adding that "those who are taught to run away from hard questions will not even make it past their first job interview."[33]

In addition to his studies and his right-wing agitation, Thiel was elected to the student government both as an undergraduate and as a law student. In his first campaign statement, as a junior, he promised to bring the disruption of an "outsider" to the organization, known as the Associated Students of Stanford University: "I have no experience in the ASSU Senate. I have no experience wasting $86,000 of student money on ASSU office renovations, helping friends pack resumes with positions in the ASSU bureaucracy, and giving them disproportionate salaries on top. As an outsider looking at the current student government, I am disgusted."[34] Thiel rose to become the chairman of the appropriations committee, where he was a stickler about the rules for disbursing money collected from student dues. At one point, Thiel tried to exclude a women's group that failed to set up a meeting with him, telling the *Stanford Daily*, "People have got to understand that they have to arrange for an interview."[35] He was overruled. In another case, he stepped in to ensure that the *Review* received student support after a different committee had voted against it. Thiel's decision to give the *Review* nearly a fifth of all the money assigned to student publications was challenged as a conflict of interest, since, of course, he had helped create the *Review*. But Thiel noted that, according to the bylaws, a conflict would only kick in if he were an officer at the *Review*, and by that time he was simply a contributor.[36]

After graduation, Thiel not surprisingly gravitated to a life in law and high finance. His career arc, which began with so much promise, met with an early failure—being passed over for a prestigious Supreme Court clerkship by two conservative justices, Antonin Scalia and Anthony Kennedy. Had he secured this ultimate distinction, Thiel reckoned, he would have had an express route to the top of the conservative legal movement. "But I didn't. At the time, I was devastated," he writes. Looking back, though, he smiles: "Had I actually clerked on the Supreme Court, I probably would have spent my entire career taking depositions or drafting other people's business deals instead of creating anything new."[37]

Even after missing out on the Supreme Court clerkship, Thiel was still heavily recruited by top corporate law firms, and he joined a prominent one in New York, Sullivan & Cromwell, as an associate. But the spell was broken. He only lasted seven months and three days, he recalls with uncanny exactitude. When he left the firm, the other young lawyers were astounded, he says, overcome by the same thought: "I had no idea it was possible to escape from Alcatraz."[38] There was another relatively brief stint, as a trader at Credit Suisse First Boston, before he started his own small fund back in California.

Then Thiel met Levchin. Though Thiel wasn't a hacker, he was an expert chess player and had reason to think a bit about computers. When he was on the Stanford chess team in the late 1980s, computers were raising their game to compete with the best. In an interview with a student reporter, Thiel was complimentary, to a point. "Although some of the computer chess programs are very impressive, they are still far from perfect," he told the reporter. "Chess computers tend to go through all of the possible moves from a given position, which makes it hard for them to look very far ahead, since the possibilities grow exponentially. Human players, on the other hand, focus on specific goals in a given position. Until the computers become goal-oriented as well, they will probably continue to lose."[39] Thiel wasn't correct about how computers would need to improve before they could beat the best human players—they succeeded through brute-force calculation—but he wasn't exactly wrong, either. For all their success in competition against the top chess masters, computers still want for the deeper intelligence that can plan and scheme in pursuit of a goal. This is the type of intelligence reserved to humans. Thiel, the nonhacker, has persisted in this view of computers as "complements for humans, not substitutes," which pretty much describes the current approach to artificial intelligence in Silicon Valley.[40]

Thiel and Levchin made a nice match, with Thiel thinking about how to harness computers to assist people and Levchin finding the most efficient ways to get computers to do what you wanted. "I'd much rather focus on building than running," Levchin said.[41] The founders shared a hiring philosophy, too. They brought in people who were comfortable with each other and saw the world in the same way. Levchin assembled his technical staff by leaning on the friends he made at the University of Illinois, while Thiel looked for employees who shared his political views by using the *Stanford Review* as a feeder organization. If D. E. Shaw & Co. was run like a computer lab focused on understanding finance, PayPal was a computer lab whose administrators happened to come from a right-wing think tank.

These two distinct pools of candidates for PayPal resulted in an initial staff that was nearly all white and nearly all male. Of the six original founders, the one nonwhite person was an immigrant from China. All were men. All but Thiel were twenty-three or younger. Four had built bombs in high school.[42] A picture of the staff six months later shows that it had grown to thirteen and included one woman, an office assistant.[43] "The early PayPal team worked well together because we were all the same kind of nerd," Thiel recalled. "We all loved science fiction: 'Cryptonomicon' [by Neal Stephenson] was required reading, and we preferred the capitalist Star Wars to the communist Star Trek."[44]

PayPal began as a company named Confinity, a combination of "confidence" and "infinity." The name was meant to convey the reliability of its initial business, which was transmitting money wirelessly between Palm Pilots, though Thiel and Levchin settled on a more promising business plan of transferring money via email and renamed the company PayPal. Only then did the libertarian rhetoric flow, with the promise of giving "citizens worldwide more direct control over their currencies than they ever had before."[45] In practice, however, PayPal never managed to escape the reach of governments, particularly the U.S. government. The company remained tightly moored by state and national regulators who, depending on the jurisdiction, might apply to PayPal the rules for banks or money transmitters.[46] No revolution was launched; no business was ever done outside of the government's watchful eyes.

What PayPal offered was an easy, secure way of transferring money via email, a convenience that could become the basis of a powerhouse company once you added a little shrewd thinking about people and markets. The key to success, Thiel and Levchin quickly concluded, was to get enough users of PayPal to convince businesses and individuals that they had to accept its payments. This was the same "network effect" sermon that Andreessen had given before there even was a Web economy. Keep adding users by any means necessary and suddenly people will be begging to join. To start, however, you must do the begging, while also employing some basic Psych 101 techniques to win over the public.

"PayPal was a very friendly name," Thiel explained. "It was the friend that helps you pay."[47] No doubt even more compelling was PayPal's plan to give users $10 for joining and another $10 for referring someone. The idea of potentially handing over $20 to whoever came knocking wasn't considered particularly radical at the time. During the dot-com boom, Levchin recalled, "the classic insanity of Silicon Valley was basically selling dollar bills for 85 cents."[48] It's hard not to compare Thiel and Levchin to scientists teaching rats how to complete a maze. Thanks to the immediate connection with its customers through the Web, PayPal could closely monitor how incentives influenced behavior, tinkering with the size of payouts for referring new customers and raising the fees charged for using a credit card to fund a PayPal account to discourage the practice.

Still, growth was slow until PayPal identified a flaw in the growing Web economy: eBay, already a hugely successful commerce site, never created an effective way for customers to transfer money electronically. Most sellers were too small to accept credit cards on their own, and, incredibly, were content to conduct business via snail mail with checks or money orders. "It was a clumsy process for an Internet service, one that PayPal could clearly improve," writes Jackson, the *Stanford Review* editor who was brought in as a PayPal executive and describes stumbling on the untapped eBay market.[49] PayPal focused laser-like on signing up the eBay clientele, knowing that each "power seller," for example, who signed on to PayPal represented a network in her own right, the center of orbit for buyers.

One tactic was to create "bots," simple computer scripts programmed by Levchin, which imitated human bidders on eBay. With these bidders, however, there would be a catch, which they explained via email. They could only complete their purchase through a new service called PayPal—maybe you want to join me there? The mysterious arrival of these eager new bidders ought to have been enough reason for sellers to sign up for PayPal, since more bidders at an auction site meant higher prices. But PayPal executives thought of yet another inducement: What if the bots told sellers that the purchases would be donated to charity? Will you sign up for PayPal now?

Pulling off this plan required more than programming skills, however; PayPal had to find a charity willing to accept sight unseen the random knickknacks the bots had bought. Jackson was given this assignment and discovered that many large charities didn't want to be part of this unusual scheme. Ultimately he found a local branch of the Red Cross that agreed to accept whatever PayPal sent them and the new charitably minded bot was then released into the eBay ecosystem.[50] With the eBay strategy in place in early 2000, PayPal began a growth spurt. The total number of users increased by more than 5 percent a day, from 100,000 in February to 1 million in mid-April; among the eBay subset, the growth rate was almost twice as fast.[51]

EBay initially didn't take these incursions by PayPal very seriously. Periodically, executives would question why they were allowing another company to build a business off of their own. They would either come up with a plan for an eBay online payment system or choose a finance company to partner with. The efforts were at best half-hearted, while PayPal, as we've seen, was always full steam ahead. Later, when eBay executives decided to drop the hammer and try to chase away PayPal by exploiting its obvious advantages with its own customers—say, by adding extra hurdles for users of PayPal—the libertarian-leaning company didn't hesitate to threaten to enlist U.S. regulators to restrain what they argued was eBay's anti-competitive behavior. Reid Hoffman, a Stanford friend of Thiel's who became PayPal's executive vice president, for a time managed to keep the path clear at eBay by merely raising the specter of government intervention.

At this point, PayPal was committed to losing millions of dollars a month to build up its audience quickly and was succeeding on both fronts. Adding to PayPal's financial challenges—beyond, ironically, its growing popularity—was an aggressive, direct competitor, x.com, which was four blocks down the road and led by another young, ambitious entrepreneur, Elon Musk. Rather than destroy themselves in competition, the two companies merged after intense negotiations. The combined company kept the PayPal service, renamed X-PayPal, as part of x.com's suite of online financial services. The company managed to raise $100 million in investment capital just days before the dot-com bubble burst in April 2000. Initially, Musk was the top dog in the merger and Thiel was pushed to the sidelines, so much so that he resigned from the company. Six months later, an executive-led coup brought Thiel back as well as the PayPal identity; Musk left day-to-day operations, but remained the largest single shareholder of the combined company with a stake of more than 10 percent.[52]

Despite the wreckage all around from the expanding dot-com collapse, PayPal appeared to be in good shape. Growth continued to be frantic and the competition had been kept at bay, either through merger or threat of government regulation. There was money in reserve. Yet PayPal still faced an existential threat in the form of consumer fraud. Solving this problem would be crucial. So crucial, in fact, that Levchin says he prefers to think of PayPal as "a security company that pretends to be a financial services company."[53]

PayPal was vulnerable to fraud by customers who, among other things, would use the service to purchase a product and then ask their credit-card company to perform a "chargeback," claiming something was amiss in the transaction. Levchin jokes about how ill prepared he and Thiel were to deal with scams like that: "Somebody told us, 'You're going to be drowned in chargebacks. You're going to die under all this massive pressure of all these people who are going to be out there just to take your money.' Peter and I were going, 'What's a chargeback? We never heard of this. OK, well, we don't have to worry about stuff we don't know.' So we just went right along. And six months into it, we still had no chargebacks. So we figured that people are actually fundamentally good. 'It's all right. No one is going to charge money back.'" Levchin then learned that six and a half months typically pass before the first chargebacks begin to appear, and sure enough, two weeks later PayPal was fast being overrun by fraud, to the tune of $12 million a month in June 2000.[54]

One response by PayPal was to become an early innovator in CAPTCHAs, tests meant to screen out malevolent bots that help in chargeback fraud. The tests involved completing a task—like copying the letters shown in an image—explicitly designed for people. In a bit of AI humor, Levchin called the technique a "reverse Turing test"; that is, instead of trying to welcome machines by offering a broad definition of intelligence, as Turing's test does, CAPTCHAs were meant to identify and keep out computers by focusing on particularly human qualities. The tests were effective in keeping out conniving computers but were useless in separating honest people from dishonest. To achieve this vital screening, Levchin would apply artificial intelligence to the information PayPal already collected about its users.[55]

The right algorithm, Levchin believed, could search through "immense quantities of behavioral data captured in processing millions of transactions per day" to detect patterns revealing fraud.[56] Think of all that PayPal knew about its customers—what kinds of items they bought, when they bought them, whom they bought them from, how they paid for their accounts. Levchin named his security program Igor in sarcastic tribute to Russian mobsters who were masters at Internet scams. Igor would have the power to freeze accounts that raised flags. Invariably, there would be "false positives," innocent people who were denied access to their own money for weeks until a human investigator could follow up, but PayPal forged ahead. A class-action lawsuit over PayPal's lack of response to consumer complaints, including over frozen accounts, was eventually settled for $9.25 million, part of the expense of creating a security system that could scale, which was a requirement because fevered, automated growth was the priority.[57]

By September 2001, PayPal had 10 million registered users.[58] The fraud rate, which at its worst amounted to well above 1 percent of the total money being exchanged in the system, was reduced by more than two-thirds because of Igor and other methods. The way was paved for PayPal's profitability and inevitable IPO. The company went public in February 2002, the first offering after the 9/11 attacks, making the founders and early employees millionaires many times over. The company hosted a party with kegs of beer and a piñata in the shape of a dollar sign, while Thiel chose to celebrate in his own special way. He played chess simultaneously against the rest of the executive staff. Only one opponent of the ten managed to beat him: Sacks, his good friend and the company's chief operating officer.[59]

Even after the IPO, PayPal was feeling the pressure. EBay again focused on controlling the online payments on its site, and PayPal responded by filing a formal complaint with the justice department and Federal Trade Commission accusing eBay of anti-competitive acts. This was an uncomfortable step for Thiel, Jackson writes: "While philosophically reluctant to get the government involved, Peter recognized that the company's increasingly unstable competitive situation, coupled with its legal and stock market woes, posed a risk too great to be ignored and consented to filing the complaint."[60] That complaint would have an unfortunate blowback not long thereafter, when PayPal agreed to be bought by eBay for $1.5 billion in August 2002.[61] If eBay could exploit anti-competitive advantages in online finance when it had a relatively strong rival like PayPal, how fair and open would the market be once it had absorbed PayPal? PayPal's own accusations against eBay should have been compelling evidence that one acquiring the other would be anti-competitive. The legal teams for both companies certainly were concerned. But the justice department under George W. Bush chose to keep its hands off.[62]

The PayPal experience was an intense, three-year ride for a relatively small team that ended in a significant cash payout: Thiel made about $55 million, while Musk left with three times as much.[63] But those events from the turn of the millennium remain influential for a number of reasons, not least for releasing into the wild a wealthy, self-satisfied Peter Thiel, who has continued practicing his unique brand of disruption up to this very day.

After PayPal, Thiel created a company, Palantir, that built on

Levchin's algorithms for analyzing and making judgments based on an individual's highly personal digital records. Named after magical stones in *The Lord of the Rings*, Palantir helps governments and private companies make judgments from online and offline records based on patterns recognized by algorithms. For example, the company produces software that in seconds can scan through hundreds of millions of pictures of license plates collected by the Northern California Regional Intelligence Center, pieces of information that can be interpreted with the help of other large data sets. Palantir's chief executive, Alex Karp, a law school friend recruited by Thiel, defends his company's role in sifting through this material, which was collected by the government, after all. "If we as a democratic society believe that license plates in public trigger Fourth Amendment protections, our product can make sure you can't cross that line," Karp said, adding: "In the real world where we work—which is never perfect—you have to have trade-offs."[64]

For someone identified as a "libertarian," Thiel has been comfortable operating businesses that relied on analyzing the personal information of its customers or the general public. Just as profiling by PayPal kept it afloat by excluding potential fraudsters, well-conceived government investigations, Thiel contends, keep America safe. After revelations by Edward Snowden about the government's surveillance capabilities, Thiel was asked if he thought the National Security Agency collected too much information about United States citizens. Thiel didn't object to those practices from a libertarian perspective but, rather, said he was offended by the agency's stupidity. "The NSA has been hoovering up all the data in the world, because it has no clue what it is doing. 'Big data' really means 'dumb data,'" he told readers of Reddit who asked him questions. "BTW, I don't agree with the libertarian description of the NSA as 'big brother.' I think Snowden revealed something that looks more like the Keystone Kops and very little like James Bond."[65]

Similar to Andreessen, Thiel lately has combined the roles of investor and public intellectual. Of Thiel's many successful investments—LinkedIn, YouTube, and Facebook come to mind—perhaps his most far-sighted has been the decision to publicly back Donald Trump for president, which required Thiel to break ranks with his Silicon Valley peers. In return for his prime-time endorsement on the final night of the Republican National Convention in Cleveland, as well as $1.25 million in contributions to Trump's campaign through affiliated super PACs and direct contributions, Thiel was rewarded with a place of privilege when president-elect Trump met with tech leaders during the transition, and an important advisory role in the new administration. Who knows what dividends are yet to be collected?[66]

The Trump endorsement reestablished Thiel's reputation as a uniquely polarizing Silicon Valley figure, a Trumpian character, you might say. Indeed, Thiel has become an almost toxic spokesman for the tech world, so much so that his close friends and business partners, like Zuckerberg and Hoffman, have felt obligated to defend their relationships publicly. During the presidential election, Zuckerberg was confronted by Facebook employees who objected to Thiel's continued role on the company's board of directors because of his support for Trump. In a fine example of rhetorical jujitsu, Zuckerberg referred to Facebook's commitment to diversity to answer those who were appalled by Trump's disparagement of Mexicans, Muslims, and women, among others, and the idea that a board member could be supporting his candidacy. "We care deeply about diversity," Zuckerberg wrote in defense of Thiel. "That's easy to say when it means standing up for ideas you agree with. It's a lot harder when it means standing up for the rights of people with different viewpoints to say what they care about. That's even more important."[67]

No doubt Thiel is an odd bird with a penchant for fringe ideas. In his pursuit of limited government, he has given substantial financial support to seasteading, which encourages political experimentation through the development of floating communities in international waters, presumably outside the reach of governments. He is unusually obsessed with his own death and sickness, a condition he traces back to the disturbing day when he was three and learned from his father that all things die, starting with the cow who gave his life for the family's leather rug.[68] Thiel supports a range of potential life-extending innovations, including cryogenics, which involves keeping a body alive by cooling it; genetic research to fight diseases; and, most resonantly, a treatment based on cycling through blood transfusions from young people in the belief that the vigor therein can be transferred to the older recipient. Thiel says he is surprised that his obsession with death

is considered weird—for what it's worth, he considers those compla-
cent about death to be psychologically troubled. "We accept that we're
all going to die, and so we don't do anything, and we think we're not
going to die anytime soon, so we don't really need to worry about it," he
told an interviewer. "We have this sort of schizophrenic combination
of acceptance and denial . . . it converges to doing nothing."[69]

Yet, cut through Thiel's eccentricities and harsh language and you
discover that Thiel is simply articulating the Know-It-All worldview
as best he knows how. In Thiel's ideas one finds Terman's insistence
that the smartest should lead, as well as his belief in using entrepreneurism and the market to introduce new technologies to the people. There is the hackers' confidence that technology will improve
society, as well as their suspicion of ignorant authorities who would
try to rein in or regulate the best and brightest. There is the successful entrepreneur's belief that the disruption that has made him
fabulously wealthy must be good for everyone. The main difference
between Thiel and his peers is that he acts forcefully and openly in
support of his ideas, while they are inclined to be more cautious and
circumspect.

As we noted above, Stanford may embrace the idea that its students
should become entrepreneurs, but only Thiel pays students to drop out
and start a business. Larry Page of Google may propose the creation of
"some safe places where we can try out some new things and figure out
what's the effect on society, what's the effect on people, without having
to deploy it into the normal world," but only Thiel backs floating sea-
based states.[70] Those peers may privately worry that democracy isn't
the ideal way to choose our leaders, but Thiel will write straightfor-
wardly in a 2009 essay for the libertarian think tank the Cato Institute
that "the vast increase in welfare beneficiaries and the extension of the
franchise to women—two constituencies that are notoriously tough for
libertarians—have rendered the notion of 'capitalist democracy' into
an oxymoron." For these reasons, Thiel names the 1920s as "the last
decade in American history during which one could be genuinely opti-
mistic about politics," though presumably 2016 restored his faith in the
electoral process.[71]

PayPal only managed to become a valuable company under Thiel's watch because eBay never could squash its tiny rival, thanks in part to the protection of the U.S. government. The decision of PayPal to complain that eBay was anti-competitive can appear hypocritical in light of Thiel's anti-government views or even in light of the company's decision to turn around and be acquired by eBay only months later. Yet when you get to brass tacks, Thiel's complaint against eBay wasn't so much about its monopoly powers, but that it was becoming a monopoly in online payments instead of PayPal. According to Thiel, a truly free market, with perfect knowledge and perfect competition, leads to failure for everyone. "Under perfect competition, in the long run *no company makes an economic profit*," he writes, adding the emphasis. "The opposite of perfect competition is monopoly." Thus, the goal of any sane start-up should be to create a monopoly.[72]

When Thiel uses the term *monopoly*, he hastens to add, he does not mean one based on illegal bullying or government favoritism. "By 'monopoly,' we mean the kind of company that's so good at what it does that no other firm can offer a close substitute," he writes in *Zero to One*, his business-advice book.[73] Yet for a company involved in online payments or for a social network like Facebook, being good at what one does is directly tied to the network effect—that is, becoming and remaining the service that is so dominant you *must* belong. Ensuring that your business has no viable competitors is at the heart of monopolistic success in social networks, a lesson that Thiel has drilled into his protégé, Marc Zuckerberg. Under Zuckerberg's leadership, Facebook has managed to keep growing and growing, spending billions to buy out any rival social networks, like Instagram and WhatsApp, before they could grow to challenge Facebook, with one notable exception—Snapchat. Founded by a pair of Stanford fraternity brothers in 2011, Snapchat rejected a reported multi-billion-dollar offer from Facebook in 2013 and has watched as Facebook aggressively copied its most popular features for sharing photographs.[74]

For Thiel, monopoly businesses like Google, Facebook, and Amazon serve as a welcome replacement for government. Freed from the unrelenting competition of the market, these businesses can afford to have enlightened values, like investing in the future or treating their employees well. They can actually think about society as a whole. Google, he writes, represents "a kind of business that's successful enough to take ethics seriously without jeopardizing its own existence. In business, money is either an important thing or it is everything." Dominant tech businesses like Google are "creative monopolies" as well, which means that they won't sit on their profits in the manner of so-called rent collectors but will push new ideas. "Creative monopolists give customers more choices by adding entirely new categories of abundance to the world," he writes. "Creative monopolies aren't just good for the rest of society; they're powerful engines for making it better."[75]

Under this theory of benevolent monopolies, government regulations and laws are unnecessary. Taxes are in effect replaced by monopoly profits—everyone pays their share to Google, Facebook, Amazon, PayPal. And in contrast to the government, these profits are allocated intelligently into research and services by brilliant, incorruptible tech leaders instead of being squandered by foolish, charismatic politicians. Levchin, during an appearance on *The Charlie Rose Show*, was asked about the libertarian cast to Silicon Valley leaders. He said he personally was OK with taxes being used to build and maintain roads, for well-functioning law enforcement and national security. For helping those less fortunate, too. But, he added, "I have relatively low trust in some of my local politicians . . . to spend my taxes on things that really do matter. And so this lack of inherent trust of the local or broader political establishment is probably the most defining, most common feature of Silicon Valley 'libertarians.'"[76]

In Thiel's version of this anti-democratic fantasy, where tech businesses set policy priorities rather than elected officials, the public need never learn the truth, that they are in essence paying "taxes" to companies while government can be belittled and whittled away. "Monopolists lie to protect themselves," Thiel writes. "They know that bragging about their great monopoly invites being audited, scrutinized, and attacked. Since they very much want their monopoly profits to continue unmolested, they tend to do whatever they can to conceal their monopoly usually by exaggerating the power of their (nonexistent) competition."[77] And the transfer is complete, from democracy to technocracy, through monopolistic tech companies that are so indispensable they impose a tax on the economy and no one complains.

This surely represents a scary political future, but it bears repeating that Thiel is no marginal character in Silicon Valley. Not only are his views surprisingly mainstream, but he operates at the very heart of the tech world as an investor and a trusted advisor to a new generation of leaders, who first spread his influence in the Valley through a network of former PayPal employees. They provided each other with cash, counsel, and contacts and called themselves, a bit facetiously, "the PayPal mafia." Their offspring include YouTube, Yelp, LinkedIn, Tesla, and, by extension, Facebook, whose first outside investment opportunity was passed from one PayPal veteran, Reid Hoffman, to another, Thiel, once Hoffman concluded that his new company, LinkedIn, could pose a conflict of interest.

In 2007, a crew of a dozen or so of these "made men" went so far as to pose for a group photo at Tosca, a San Francisco café, garbed in cliché Italian mafia outfits. That photograph, for an article in *Fortune* magazine, quickly joined the annals of over-the-top Silicon Valley images, right up there with the *Time* cover a decade earlier that featured a barefoot twenty-four-year-old Marc Andreessen sitting on a throne next to the headline, "The Golden Geeks."[78] Levchin is in the front, wearing a black leather jacket; Hoffman sports an open-collared silk shirt revealing a gold chain; others donned track suits. Front and center is Thiel in a dark, pinstriped suit, purple shirt and tie, and pinky ring.[79]

8. REID HOFFMAN ET AL.

_____

"My membership in a notable corporate alumni
group in Silicon Valley has opened the door . . ."

The Silicon Valley era that produced the PayPal mafia companies—the years immediately following the tech stock market crash of 2000—couldn't have been more different from Andreessen's golden age. This bleak time, which was exacerbated by the 9/11 attacks, has been called the valley's "nuclear winter" when high-profile dot-coms disappeared and weren't replaced. Over a period of three years, venture capitalists cut their investments by 80 percent.[1] The events called to mind nothing so much as when a massive meteor collided into the Earth eons ago, casting a pall that wiped out the large, lumbering dinosaurs. We would do well to recall how that ancient calamity provided a great opportunity to the early mammals, our ancient ancestors. Once a minor class of animals on the fringe, the mammals were finally out of the dinosaurs' shadow and could use their bigger brains and more advanced social skills to take charge of a relatively barren ecosystem. In due course, their descendants—us, of course—reached the top of the food chain. In much the same way, Silicon Valley's nuclear winter provided a clear path to dominance for a group of clever, well-connected, already wealthy entrepreneurs.

Tough times do have a way of clarifying things for wannabe entrepreneurs. Remember, it was only during the tech world's nuclear winter that Google's investors put their feet down and insisted that the company's founders, Sergey Brin and Larry Page, find a way to turn their happy, re-created computer lab into a profitable business. By 2001, barely a year later, Google was profitable; by 2004, it was a runaway success. PayPal, too, had been forced by the tough economic conditions to focus on a winning strategy; using venture capital to purchase an audience $20 at a time was no longer an option. Strange to say, but this actually was an extraordinary time to start a tech business, assuming you had the resources to get off the ground. Was the Web any less useful after the bubble in tech stocks burst? Of course not. The Internet was still going strong, even if many Internet companies weren't.

In that fateful year, 2000, about 120 million Americans had access to the Internet in their homes; by 2003, the total had grown by nearly 60 million, to 179 million.[2] The global growth in Internet users over that same period was even more impressive: roughly 415 million users in 2000 became more than 781 million by the end of 2003.[3] Online commerce kept growing during these dark days as well. Total e-commerce in the United States in 2000, according to Census data, was $27.6 billion. That total doubled by 2003, to $57.6 billion.[4] As far as anyone could tell, the winning Web strategy remained largely the same, too: be quick to market and ruthlessly exploit the network effect, the virtuous loop that drives people to join a service because others have already joined.

Start-ups of this era would need to run on lean budgets, but then again, conditions had never been better for pulling that off. To begin with, there was the relative lack of competition, which meant that a new company could build an audience without fear of being undercut by a well-financed rival. Another advantage was that expenses were falling across the board—computer hardware was getting faster and less expensive, while free software was becoming more reliable and pervasive. Perhaps the greatest advantage of that era was the spread of faster broadband Internet connections, which enabled enthusiastic Web users to post online all manner of creative work and personal information—essays, recommendations, technical advice, photos, videos.

The tools for collaboration and direct publishing that Tim Berners-Lee had argued for at the start were introduced to the mainstream during these years as Web 2.0. This phrase immediately stuck in his craw, however. What exactly was new—2.0-ish, that is—about these companies built around user contributions? They were simply fulfilling the original vision for the Web, where "interaction between people is really what the Web is."[5] There was one vital difference, however. In Web 2.0, the tools for sharing and publishing were part of a social network or service that profited off of what you created there. You weren't publishing a video to the Web, you were publishing a YouTube video; you weren't writing a restaurant review on the Web, you were posting a Yelp review. The truly new idea expressed by Web 2.0 was a commercial one, and perhaps that is why Berners-Lee couldn't recognize it.

Directly profiting from what the public posted online certainly was a much better deal for tech businesses. While Google had proven that there was a business to be made by treating everything freely published on the Web as a collection of "user-generated content," think how hard the task was: Google had to collect and store enormous quantities of data and then build a complicated algorithm to extract value from it. By contrast, a Web 2.0 company could find itself flooded with valuable contributions from its own users, which it could exploit. The method for profiting might have been the same—publishing compelling material that would appear next to relevant advertisements—but a Web 1.0 company had the enormous challenge of sifting a fast-changing river of information for gold powder, while the others were, in effect, melting down gold jewelry that had arrived as a package left on the porch. Google eventually understood this distinction, too, and began collecting troves of user-generated content that it could exploit, whether through homegrown services like Gmail or Google+ or through its acquisition of businesses like YouTube or its ambitious scanning of the world's libraries.

A trailblazer of Web 2.0 strategy was HotorNot.com, which was founded in 2000 and immediately became an online sensation by allowing a user to upload a photo and receive a snap judgment on "hotness" from the site's visitors, who voted on a scale of 1 to 10. When three PayPal veterans—two programmers, Jawed Karim and Steve Chen and a designer, Chad Hurley—created YouTube, they were intending to bring the hot-or-not idea to video. In an interview, Karim said he was impressed by HotorNot because "it was the first time that someone had designed a Web site where anyone could upload content that everyone else could view. That was a new concept because, up until that point, it was always the people who owned the Web site who would provide the content." After a few months, YouTube broadened its mission by allowing all manner of online videos and was acquired by Google in 2006 for $1.65 billion.[6]

This openness to user contributions on the part of Web 2.0 companies easily blended into a laid-back approach to potential copyright violations, in that much of what users wanted to "share" was material others had made or owned. This was the hacker ethic from the computer labs, which insisted that a frictionless system for passing along material meant that anyone should be allowed to do so. Copyright owners in the music, film, news, and TV industries, however, quickly recognized that these sites could threaten their businesses. Silicon Valley would have to fight in federal and state courthouses for greater tolerance for posting copyrighted material.

Google's YouTube division, among others, came to rely on the "safe harbor" provision of the 1998 Digital Millennium Copyright Act to be spared liability for hosting illegal copies of copyrighted material. As long as they promptly removed material when notified by the copyright holders, these services wouldn't be liable for what might otherwise be infringing behavior. YouTube benefited from being a comprehensive video-sharing network with many uses beyond transmitting copyrighted material. Courts treated it differently from, say, Napster, the peer-to-peer music sharing site that was seen by friends and foes alike as principally a means of acquiring copies of copyrighted music; Napster was shut down by a judge's order in July 2001.[7] YouTube thrives to this day. The safe harbor provision proved crucial to the acceptance of Web 2.0 sites, helping to mainstream the practice of copying and posting clips of copyrighted material, even as these services created automated systems for identifying and removing material once copyright owners complained.[8]

Thus the Web, even through 2.0, made another detour from the Berners-Lee model of personal and direct online collaboration. Instead, collaboration continued to be centralized and commercialized, as Andreessen had conceived from the start, like a global TV network. YouTube and Facebook and Amazon became the way people accessed news and entertainment. The single-minded pursuit of large online audiences by tech companies threatened the financial under-pinning of traditional news outlets and cultural businesses and orga-nizations, in yet another example of how the combination of hacker freedoms and entrepreneurial greed produced the social disruption we are experiencing today.[9]

The formula for Web 2.0 success at the dawn of the new millennium was ideally suited to members of the PayPal mafia: there was a chance for great success quickly, provided that you arrived on the scene with a new service that was interesting, reliable, and scalable. During the nuclear winter in Silicon Valley, when outside investor money had largely disappeared, the PayPal mafia had each other—millionaires who either were talented programmers or knew how to run a start-up. They established the kind of business-friendly support system that Terman perfected at Stanford in the 1950s and '60s: there were early investments that acted like government grants; access to well-trained former colleagues, who were the equivalent of graduate students or research assistants; and there often was spare office space to borrow, so a team could move in immediately.

In addition to YouTube there was Slide, a photo-sharing service from Max Levchin, which was quickly snapped up by Google for $200 mil-lion; Yammer, cofounded by David Sacks, which produced an internal communication tool and was bought by Microsoft for $1.2 billion in 2012; and Yelp, the publicly traded company cofounded by two former PayPal programmers, which harnesses user responses to create reviews of a restaurant or even of a local doctor. Facebook, which launched in 2004, joined the PayPal mafia's orbit when its founder, Mark Zucker-berg, moved to Silicon Valley and needed an infusion of cash.

To see how the PayPal mafia worked in practice, consider how Reid Hoffman used the connections he made at PayPal to build his start-up, LinkedIn.[10] In 1997, Hoffman cofounded the Web site Social-Net, which tried to turn online connections into offline relationships, romantic and otherwise. Looking back, you could say that SocialNet was too early to social networking, and consequently never managed to build a large audience or figure out how to profit from the small one it had. When Hoffman sold the company, he had a mere $40,000 to show for his efforts.[11] After giving up on SocialNet, however, Hoffman was invited by Thiel, his close college friend, to join PayPal, first as a board member and later as a top executive.[12] At PayPal, as we've seen, Hoffman was front and center in maintaining the company's access to eBay's customers one way or another; at times, he would cajole eBay's lawyers into playing nice, at others he would encourage the Department of Justice to investigate eBay for anti-competitive behavior.

Once eBay decided to buy its pesky rival for $1.5 billion, Hoffman's profile in Silicon Valley changed radically. "They tracked SocialNet as modestly interesting," Hoffman said of his fellow entrepreneurs and investors, but, he added, "PayPal was my induction into the circuit. It pegged me as legit; it pegged me as a player."[13] And, indeed, when Hoffman proposed a new company called LinkedIn for building professional networks, he was able to get started quickly because of his enhanced reputation and personal wealth, as well as a network of friends who served as cofounders, early employees, and investors. Financing for the company came from, among others, Thiel and Keith Rabois and other multimillionaires created by PayPal's sale to eBay; the company's first office space was provided by a former PayPal colleague. Later, once LinkedIn was a viable company, Hoffman was able to pay it forward, giving office space to the former PayPal employees who created YouTube and joining them as an early investor.[14]

There is something jarring about a group of self-styled survival-of-the-fittest free-marketeers committing to a strategy of collective risk and mutual support. At least one pillar of the Silicon Valley ideology was toppled by this arrangement: that success was handed out to an entrepreneur strictly according to ability and hard work, no matter his station in life or place of origin. Marc Andreessen once expressed this faith in individual merit in an interview with the *New York Times* journalist Tom Friedman, offering another example of how the world is flat: "The most profound thing to me is the fact that a 14-year-old in Romania or Bangalore or the Soviet Union or Vietnam has all the information, all the tools, all the software easily available to apply knowledge however they want. That is, I am sure the next Napster is going to come out of left field."[15] Somehow, things haven't quite worked out that way. Instead, a collection of male executives from a single company, a few of whom were hired as much for their right-wing political beliefs as for any latent computer or business talent, managed to create a roster of successful companies. A kid in Romania, it seems, no matter how talented, didn't stand a chance against these guys.

Considering their origin in university friendships and earlier start-ups, networks like the PayPal mafia tended to be nearly uniform when it came to sex, race, and educational background: white, male, elite. Somehow, though, the experience of profiting from connections and college friendship hasn't diminished the lecturing from Silicon Valley about how other institutions—typically highly unionized ones like the public school system or the automobile industry—are rife with favoritism. Here is Hoffman explaining Detroit's decline in *The Start-Up of You*, his business-advice book: "The overriding problem was this: The auto industry got too comfortable. . . . Instead of rewarding the best people in the organization and firing the worst, they promoted on the basis of longevity and nepotism."[16] Hoffman makes no mention of any similarity to the PayPal mafia, which he describes this way: "My membership in a notable corporate alumni group in Silicon Valley has opened the door to a number of breakout opportunities."[17]

Though considered a liberal in Silicon Valley, Hoffman adheres to the consensus view there that society is organized around unbridled competition within a market. "Keep in mind," he writes, "that the 'market' is not an abstract thing. It consists of the people who make decisions that affect you and whose needs you must serve: your boss, your coworkers, your clients, your direct reports and others. How badly do they need what you have to offer, and if they need it, do you offer value that's better than the competition?"[18] And you had better please this market, he continues, because the social safety net is an illusion. In a footnote, Hoffman counsels his readers to "consider the Social Security tax that comes out of your paycheck like you would a loan to a second cousin who has a drug problem. You might get paid back, but don't count on it."[19]

Thiel and Hoffman say they became friends through dorm-room political arguments including one over Thiel's agreement with Margaret Thatcher's statement that "There is no such thing as society. There are individual men and women and there are families." Hoffman considers himself on the left, and vehemently disagreed with Thiel over this harsh view of humanity.[20] But if the society that Hoffman advocates for is little more than a marketplace where individuals must slug it out, then the gap between the two isn't so significant. Hoffman's society—Silicon Valley's, really—ain't much of a society.

For instance, knowing how unforgiving the market is, why would Hoffman begrudge workers the right to form a union as a way of ensuring some stability and protection? Hoffman prefers that each of us build "a network of alliances to help you with intelligence, resources and collective action."[21] A wealthy Stanford graduate, Hoffman has undoubtedly benefited greatly from his "mafia" of similarly situated peers. However, an economy based on personal networks undoubtedly disadvantages nearly everyone else, even ambitious, well-educated entrepreneurs like Jimmy Wales, born in Huntsville, Alabama, and eager to cash in on the Web craze.

9. JIMMY WALES

"Wikipedia is something special"

I n the mid-1990s, Jimmy Wales had a lot in common with Jeff Bezos and Peter Thiel—an early career in finance, libertarian-leaning politics, youthful fascination with science and computers, and a determination to make it rich from the Web. What he lacked, however, was their elite educational pedigree and Wall Street experience. And that would make all the difference for Wikipedia, the brilliant project Wales has shepherded since its inception in 2001.

The encyclopedia anyone can edit, Wikipedia appeared to offer a golden ticket to a Web 2.0 entrepreneur like Wales with its oodles of user-generated content. But Wikipedia has taken a different path: no advertising appears next to the articles its editors have written or the photos they have posted; there are no suggestions of products to buy based on your search; the project doesn't track its visitors, and hasn't once thought to acquire a rival online encyclopedia. Wikipedia is a shining outlier in the commercialized Web, a site that has benefited enormously from the network effect but has never sought to profit from it. And Wales, the Web entrepreneur, has watched it all happen.

Jimmy Wales was a precocious kid, born in Huntsville, Alabama, in 1966. His early education came at a small school started by his mother and grandmother, though he later attended a private high school.

Huntsville was a scientific hub, the home to a space research center led by the infamous Wernher von Braun, who had masterminded the Nazis' rocket program. In high school, Wales learned to program a PDP-11 minicomputer, one of the Digital machines that were beloved by the early hackers. Wales kept up his interest in computers and programming as a side hobby while he studied finance at Auburn and the University of Alabama, getting bachelor's and master's degrees. He left a PhD program in finance at Indiana University to become a futures and options trader in Chicago.[1]

The Netscape IPO in late 1995 caught Wales's attention, and he made another career change: Web entrepreneur. The next year, Wales started his own online business called Bomis, which included a simple search engine and an index of Web sites assembled by registered users, who were called "ringmasters." What these human indexers produced were "rings" about a specific topic, one site joining the next. In classic Web 2.0 fashion, Bomis was planning to profit from the indexes it solicited by surrounding them with advertisements. The company made an appeal to ego and self-interest to spur contributions. "If your ring is really good, we may choose to include it in our search engine and tree structure," the Web site said. "This is a great honor because it means that we think you've gotten at the essentials of some topic, and that we think you have been reasonably fair and unbiased in your selection and descriptions. (Of course, putting your own page first in the ring is PERFECTLY fine with us!)"[2] This was not an ideal situation for runaway growth, however; Bomis was highly dependent on the work of humans, either its small group of contributors or its even smaller staff. There was little artificially intelligent about the company, and thus little chance of its scaling.

Wales was open to any new idea for encouraging users to produce material that he could then sell ads against; like many sites of this era, Bomis wasn't shy about helping its users find porn.[3] In January 2000, when the company had about a dozen employees and tech stocks were at their peak, Wales proposed that Bomis publish an online encyclopedia created by users, which he called Nupedia. For the first editor of Nupedia, Wales hired Larry Sanger, a technically proficient philosophy PhD, whom Wales had met through an online discussion group about Ayn Rand, the "virtue of selfishness" philosopher.[4] As any sensible encyclopedia operators would do at that time, Sanger and Wales established a thorough review system for articles, one so thorough that after a year barely twenty articles had been given the green light to be published. At that rate, Nupedia would challenge the *Encyclopedia Britannica* in a few thousand years.[5]

This system obviously couldn't last, and the dot-com crash, again, had a way of bringing clarity to the situation. Bomis hadn't discovered an automated way to unlock profits from the Web, as Google or PayPal or Amazon had. Wales and his partners had no rich friends to lean on in tough times, either. His company, which was based in San Diego and then Florida, was "on the periphery of the Internet," recalled Terry Foote, an old friend of Wales's who was Bomis's advertising director. "Jimmy and I, each in our own different ways, pounded on the metaphorical doors of all the Silicon Valley big shots, and mostly what we got was a solid wall of silence."[6] Under pressure to rethink Nupedia, Sanger recalled proposing a way to speed up the review process, but "everything required extra programming," he said. "By then, Jimmy Wales was worried about keeping costs low, the dot-com boom was turning to bust, and so whatever we did to solve the problem had to involve no more programming." That's where "wikis" came in. A friend told Sanger about the free wiki software that allowed users to collaborate online on a single article, roughly along the lines that Berners-Lee had planned for the original Web browser.[7]

The Web site Wikipedia.com began operating on January 15, 2001, and, like Mosaic and Google before it, found an audience almost immediately. With its software innovation, Wikipedia had six thousand entries after six months. By the end of the year, there were twenty thousand articles, most in English but hundreds of others in more than a dozen languages, including German, Spanish, Polish, and the artificial tongue Esperanto.[8] When the original wiki software was unable to handle this growth, a German volunteer created the more capable version that is still used today.

In addition to its enthusiastic volunteers, many of whom were originally drawn to Nupedia, the Wikipedia project found crucial assistance from the new Google search engine, whose algorithm didn't care a whit about offline reputation. Google knew only what could be conveyed through Web links, and according to that calculus Wikipedia could have a loftier reputation than the *Encyclopedia Britannica*, whose articles didn't freely circulate online. Speaking to a Stanford class almost exactly a year after Wikipedia was created, Sanger described the positive feedback loop his project was experiencing through Google. "We write a thousand articles; Google spiders them and sends some traffic to those pages," he said. "Some small percentage of that traffic becomes Wikipedia contributors, increasing our contributor base. The enlarged contributor base then writes another two thousand articles, which Google dutifully spiders, and then we receive an even larger influx of traffic. All the while, no doubt in part due to links to our articles from Google, an increasing number of other websites link to Wikipedia, increasing the standing of Wikipedia pages in Google results."[9]

Those first twelve months, shattered by the 9/11 attacks, included the absolute trough of the dot-com collapse. Under those conditions, Bomis had yet to make money from its runaway hit, Wikipedia. The company laid off half of the staff, including Sanger in February 2002.[10] Without Sanger's salary, the costs of running Wikipedia were surprisingly low—by 2004, total expenses for a significantly larger Wikipedia were still less than $25,000[11]—but even those sums were a drain on Bomis when it could least afford one. Sanger understandably agitated for Bomis to raise some money to restore his salary as the editor in chief of Nupedia and "chief organizer" of Wikipedia, jobs he described as the best he'd ever had. In a letter to the Wikipedia community explaining that he would have to reduce his involvement in the projects while he looked for full-time employment, Sanger opened Pandora's box, writing hopefully that "Bomis might well start selling ads on Wikipedia sometime within the next few months, and revenue from those ads might make it possible for me to come back to my old job."[12]

The mere mention of advertising was so disturbing to the volunteers who were building Wikipedia that the project was nearly destroyed by the backlash. Why this group of content generators—as opposed to people listing their résumés or reviewing restaurants—would be particularly offended by advertising is hard to explain. Was their anger based on the fact that writing an encyclopedia article is fundamentally altruistic and thus shouldn't be exploited for profit? Maybe it was Wales's laid-back posture toward the project that gave them such self-confidence? He didn't come off as a typical Silicon Valley entrepreneur, convinced how brilliant he was, treating users like idiots to be exploited. Bomis knew it was bringing very little to the table when it came to Wikipedia—the inmates were running the asylum, and that seemed to be working out fine. Perhaps the best explanation is that Wikipedia didn't begin with ads, which meant that they would have to be introduced, which gave the opposition something to fight against.

There was a final factor: Wikipedia was created under a so-called "free" software license, and those terms applied not only to the software but to the articles as well. While there certainly was no rule against running advertising next to material made under a free license—free software's most prominent activist, Richard Stallman, is fond of saying, "free as in free speech, not free beer"—this license gave anyone the right to copy all the material that appeared on Wikipedia and begin his own project, a process known as "forking." An exact duplicate of Wikipedia—a fork—could appear under a new name at a new Web address; the only condition was that the new project operate under the same free license. Immediately after Sanger made his idle suggestion about advertising, a group of contributors to the Spanish version of Wikipedia forked the encyclopedia. The articles were copied and stored on servers at the University of Seville, where the project was renamed Enciclopedia Libre Universal en Español.

The Spanish contributors were on the left politically and already annoyed to be part of a for-profit business. Furthermore, Edgar Enyedy, one of the leaders of the Spanish fork, resented having to go through Bomis to make improvements to his own community's Wikipedia. "We were all working for free in a dot-com with no access to the servers, no mirrors, no software updates, no downloadable database, and no way to set up the wiki itself," he said. "Finally, came the possibility of incorporating advertising, so we left. It couldn't be any other way." The threat that other foreign-language Wikipedias might fork if there were ads—or potentially that the English Wikipedia itself would—was enough to get Wales to take notice and say no to advertising.[13]

With the best opportunity for profiting from an online encyclopedia cut off, Wales in 2003 shifted ownership of Wikipedia to a charitable foundation; at least the project wouldn't be a financial drain anymore. A few prominent articles in the tech press spread the word about the new Wikimedia Foundation and donations from individuals and institutions almost immediately managed to cover whatever expenses were connected to running the site. In 2004, when English Wikipedia had 150,000 articles, the foundation raised more than $80,000 to cover $23,000 in expenses.[14]

Late in that year, Wales channeled his thwarted entrepreneurial energy into Wikia, a for-profit venture built on wiki principles. The articles on the Wikia site are like Wikipedia articles, created and edited by visitors, but they reflect the enthusiasm of a 'zine rather than the neutral view of an encyclopedist, which is why the San Francisco–based company recently renamed its Web site "Fandom powered by Wikia." The company has followed a traditional strategy for a Web 2.0 company, running targeted advertisements on article pages and seeking investments from venture capital firms, as well as Amazon.[15]

To this day, the Wikimedia Foundation is never short of funds, even as the size of its staff and the number of articles within its dozen or so wiki projects has grown substantially. Wikipedia alone has 40 million articles, which can appear in any of 290 different languages. In the 2016 fiscal year, expenses at the foundation were $65 million, including paying for a staff of more than 280 people. In that period, total revenue at the foundation was more than $80 million, with $77.7 million coming from donations and contributions.[16]

The unconventional development of Wikipedia presents an interesting contrast to how Google grew. Google's founders, as we've seen, were academics fundamentally opposed to advertising, who eventually succumbed to the start-up-friendly vortex surrounding Stanford. As a result, Larry Page and Sergey Brin had the resources to continue operating during dark times until they discovered a way to make advertising work. Jimmy Wales, on the other hand, was committed to Web advertising from the start and would have loved nothing better than to bring advertising to Nupedia and its new incarnation, Wikipedia. He has spent his entire entrepreneurial life trying to create user-generated content to place ads against, yet when he finally hit a gusher he was unable to capitalize.

Had Wales not stumbled on the idea of Wikipedia in absolutely the worst time for dot-coms—or if he had a network of Stanford friends like the PayPal mafia to call upon—perhaps he could have pushed through and operated Wikipedia as a business. He could have ignored the Spanish fork and introduced advertising. With the appropriate resources, he then could have hired programmers to add artificially intelligent features to make the Bomis version of Wikipedia—ads and all—such a superior experience to a forked, ad-free version that few would turn away.

Wales would occasionally return to the idea that the site might accept ads, if not for company profits, then to benefit a charity. "That money could be used to fund books and media centers in the developing world," he said in 2004. "Some of it could be used to purchase additional hardware, some could be used to support the development of free software that we use in our mission. The question that we may have to ask ourselves, from the comfort of our relatively wealthy Internet-connected world, is whether our discomfort and distaste for advertising intruding on the purity of Wikipedia is more important than that mission."[17] After the community again objected to such a flirtation with advertising, Wales gave up on the idea forever. Not that advertising is evil, he wrote later: "But it doesn't belong here. Not in Wikipedia. Wikipedia is something special. It is like a library or a public park. It is like a temple for the mind. It is a place we can all go to think, to learn, to share our knowledge with others."

In that same statement, he glosses over the early, contentious history of the project, declaring, "When I founded Wikipedia, I could have made it into a for-profit company with advertising banners, but I decided to do something different."[18] The narrative that Wales offers in regard to Wikipedia, of being tempted by, and rejecting, a lucrative opportunity to run advertisements is actually closer to what happened at Craigslist. In late 1997, you will recall, the site's founder, Craig Newmark, was approached about accepting banner ads for a Microsoft service on his site, which had already gone through a growth spurt. He turned the offer down, knowing he had "stepped away from a huge amount of money," because it wasn't in the best interest of Craigslist's users.

Wikipedia and Craigslist are indeed the exceptions that help define what is standard Silicon Valley behavior. They both have scaled to become worldwide phenomena, but not by automating every experience through artificial intelligence. Newmark welcomes personal interactions with his users and proudly describes himself as a customer service representative for the site, albeit one who also relies on automated tools.[19] Wikipedia likewise has teams of volunteers who help new users and watch out for vandalism, also aided by automated tools. Truly, they are what Berners-Lee expected the Web to be from its inception, before it took a sharp turn toward commerce. They are decentralized and collaborative; regular folks are given an important voice. Doubtful we will see their like again.

Starting in the 2000s, Thiel, Hoffman, Andreessen, and their ilk shifted from cashing checks to writing them. In their role as investors and mentors, they made sure that the next generation wouldn't make the same mistakes they had. They warned founders to keep control, pushing back meddling VCs by being sure to retain a majority of voting shares. And they advised start-ups to be ruthless in acquiring market share and then protecting that share—Thiel's stealth monopolist strategy. When Mark Zuckerberg visited the Bay Area the summer after his sophomore year at Harvard, this crew found their prodigy, like the weathered boxing trainer Cus D'Amato being introduced to the thirteen-year-old Mike Tyson, whom he would shape into a fearsome champion. Zuckerberg was a talented programmer with both a hacker's belief that computers could save the world and a Thielian ruthlessness about using the network effect to replace the Web with his own service, which would soon be renamed Facebook.

10. MARK ZUCKERBERG

"Nerds win"

The first significant computer program young Mark Zuckerberg wrote was a version of the board game Risk set in the Roman Empire. "You played against Julius Caesar. He was good, and I was never able to win," Zuckerberg recalled with a twinge of pride.[1] That's the thing about playing a video game of your own creation while barely a teenager: even if the character you control loses, you've still won. Caesar's victories, the computer's victories, were Zuckerberg's, too. This veneer of ultimate control, according to the computer science pioneer Joseph Weizenbaum, was the best explanation of what motivated the young hackers he found toiling away at all hours of the night at MIT. "No playwright, no stage director, no emperor, however powerful, has ever exercised such absolute authority to arrange a stage or a field of battle and to command such unswervingly dutiful actors or troops," he wrote back in the 1970s. "The computer programmer is a creator of universes for which he alone is the lawgiver."[2]

Zuckerberg was given the chance to build his own universes beginning in the sixth grade. This was the mid-1990s, the World Wide Web was catching fire, and PCs were well within reach for a well-to-do family like the Zuckerbergs. Young Mark wouldn't need to hustle for computer time the way young Bill (Gates) did twenty-five years earlier, proposing school rummage sales and then creating businesses just to get his hands on a minicomputer for a few hours. Mark's parents were enthusiastic partners in his youthful computer obsession, buying him a first personal computer when he was eleven; and after their son was initially stumped by a programming book "for dummies," they provided him with a tutor, as well.[3] Mark was hooked. "I'd go to school and I'd go to class and come home and the way I'd think about it would be, 'I have, you know, five whole hours to just sit and play on my computer and write software,'" he recalled about his time growing up in Dobbs Ferry, a suburb just outside New York City. "And then Friday afternoon would come along and it would be like, 'Okay, now I have two whole days to sit and write software.' This is amazing."[4]

Even as a youngster, Zuckerberg planned on shaking up the world with his brilliant programs in a way the original hackers never could have imagined. The hackers of earlier generations were basically coding to impress their computers and themselves, while Zuckerberg was coding to impress his classmates. Video games of all stripes sprang from young Mark's fingers, often based on drawings from friends who would sketch while he was programming. In 1996, he hacked together a messaging system for his dad's dental office to announce that a patient had arrived. Later, while away in prep school, Zuckerberg and a classmate, Adam D'Angelo, created an "artificial intelligence" program called Synapse that studied your music listening habits to suggest the next track to play. The program was powered by an algorithm they named "the brain," whose accuracy in assessing your musical preferences, Zuckerberg and D'Angelo insisted, was within a hundredth of a percent, whatever that means.[5]

Throughout these experiences, Zuckerberg wasn't just coding, but "building things," as his older self likes to say. Synapse gave Zuckerberg his first taste of the rewards to come; the program received a positive mention on the influential Web site Slashdot and drew the interest of a number of big companies, including Microsoft and AOL.[6] Initially, Zuckerberg and D'Angelo were reluctant to sell their program, but by the time they had a change of heart the offers (said to be in the $1 million to $2 million range) were gone. All that remained for Zuckerberg, who attended Harvard, and D'Angelo, the more talented programmer who went to Caltech, was a memorable lesson about striking while the iron is hot.[7]

Zuckerberg, with intense eyes and an easy grin, long resisted being defined by the traits of a traditional hacker, even if he put in similarly obsessive hours in front of his computer and was known to indulge in the same silly programmer jokes, like how to instruct a computer to get drunk. The hackers were antisocial and anarchic, their vision shrunken to fit the dimensions of a metal box; Zuckerberg was worldly, as proud of his broader interests—whether in human psychology or fencing or ancient history—as of his programming skills. The hackers were inward looking; Zuckerberg was constantly taking stock of his environment. The hackers considered Microsoft the Evil Empire; Zuckerberg was in awe.

The hackers loathed Microsoft not only for its opposition to the free software movement but also for what one might call aesthetic reasons. Microsoft software was plain and uninspired, as if the company wouldn't waste the effort on elegant programming since whatever it produced was destined to become the standard anyway. To hackers, the only reason to program was to do it elegantly. Zuckerberg, by contrast, told the world how impressed he was by Microsoft's commercially dominant operating systems. "I mean, I grew up using Windows 3.1 and then Windows 95, and I just thought that those were, like, the most unbelievable things," he said in a talk with Paul Graham, a self-proclaimed hacker and important investor in start-ups through Y Combinator. Graham appears taken aback by this praise of Microsoft products and replies cryptically, "In a sense, they are [unbelievable]." Zuckerberg carries on: "Yeah, they were. They really were awesome. Right?" The crowd chuckles, assuming Zuckerberg is being sarcastic. "Well, I don't know if you meant that positively, but I did. And I thought, you know, building this ecosystem was really neat, and that kind of inspired me."[8]

Yes, the hackers may have correctly observed that Microsoft products impose a dull uniformity, but Zuckerberg was willing to let that pass. He was focused on what that simplicity and uniformity provided in return: millions of people using the same operating system and software as they composed their letters or balanced their financial books or played their video games. Here was the age-old tension between indulging individuality (hackers) and imposing order so things could run smoothly (builders). Monty Python's *Life of Brian* captured it well when Reg, the ornery leader of a small sect of Judean rebels resisting Roman occupation, tries to rally his followers by asking, "What have the Romans ever done for us?" Annoyingly, they offer example after example, until Reg shouts in frustration: "All right, but apart from the sanitation, the medicine, education, wine, public order, irrigation, roads, the fresh-water system, and public health, what have the Romans ever done for us?"

Zuckerberg found a role model in Gates, the Harvard dropout who became a tech billionaire. Gates, like Zuckerberg, was a computer obsessive intent on building things—a company, operating systems, suites of software, an entire computer ecosystem. Gates, like Zuckerberg, was out of step with his fellow hackers, which became apparent when Gates, at age nineteen, accused a group of Bay Area hackers of stealing the software his company, Micro-Soft, had produced. In 2004, Gates was invited to speak to Zuckerberg's computer class, where he encouraged students to take time off to start a new project. As Zuckerberg recalled it, Gates said, "You know, one of the things about Harvard is they let you take off as much time as you want and then you can always come back, so you know if Microsoft ever falls through, I'm going back to Harvard."[9] The class chuckled. Turns out, that would be Zuckerberg's last semester before relocating to the Bay Area and (so far) never returning.

Zuckerberg is Gatesian in seeing how computing might prosper under one company's Roman-like system for connecting people across the globe. The hacker types, like Reg, would hate that rules were being imposed from above, but the public would love the ease of use that such uniformity would deliver. When Zuckerberg was still in college, he allowed his mind to wander. "I was working on this Facebook thing and I thought it would be cool for Harvard and . . . I thought that over time someone would definitely go build this version of this for the world but it wasn't going to be us." The job of creating a platform to connect the world seemed too monumental for a bunch of undergraduates to pull off. "It was going to be you know, Microsoft or you know someone who builds software for hundreds of millions of people."[10] To Zuckerberg's eternal surprise, or so he says, the task of creating a global social network fell to a bunch of kids barely out of their teens. They would be the new Microsoft and so much more.

While at Harvard, even more than in prep school or back in Dobbs Ferry, Zuckerberg had the freedom to program all the time. He routinely skipped classes and ignored study periods to tinker on his computer. Yet in college there was an opposite pull, too—opportunities seemed to be opening up before his eyes, and Zuckerberg was naturally curious about how he would fit in. Would he be liked? Would he have friends? Would he be successful? What would he make of himself? In the span of a year, Zuckerberg built a series of campus-wide Web sites, one more famous than the next, that were fundamentally social in nature. Zuckerberg was keenly aware of belonging to a community, which by turns he wanted to impress, intimidate, imitate, get to know.

First, he created a Web site called CourseMatch, which allowed students to share what classes they were taking. He was planning his schedule and wondered what students he knew from computer science classes would be studying. Was he looking for new areas of study or wanting to keep tabs on what everyone else was doing? A bit of both? In a matter of weeks, two thousand Harvard undergraduates had joined CourseMatch, which Zuckerberg ran from a laptop in his dorm room. He wasn't expecting this level of interest and his laptop was fried in the process. Zuckerberg quickly abandoned CourseMatch, after gaining valuable experience in how to create a campus-wide Web site that spread virally, and perhaps how to keep one running in the future, as well.[11]

In that same vein, Zuckerberg in the fall of 2003, his sophomore year, built the infamous Facemash, which was an even quicker hit on campus. The idea was to let students compare the attractiveness of their classmates head-to-head, the tried-and-true hot-or-not method for reaching an online audience with advertising. But Zuckerberg at this point wasn't an entrepreneur on the make. He says he wasn't planning to turn Facemash into a business and he wasn't trying to persuade students to upload personal material Web 2.0 style. In fact, much of the fun for Zuckerberg was "liberating" photos of Harvard students by hacking the university's servers. In a digital journal that appeared on the Facemash site, Zuckerberg detailed how he obtained the pictures the site used. It's a brief glimpse into his puzzle-solving mind, as each dorm on campus offered its own obstacles: for some, Zuckerberg had to deduce the student passwords before accessing a photo file, in others he only had to run a computer script that could copy photos twenty at a time. When one cache of photos proved particularly hard to hack, he writes, he took a break and opened another Beck's.[12]

After he had collected the photos and completed the programming during a couple of all-nighters, Zuckerberg published the site, which randomly matched two photos to be compared by a visitor to the site. His slogan: "Were we let in for our looks? No. Will we be judged on them? Yes." Word spread over a single weekend and there were already 450 visitors who voted at least 22,000 times. Student organizations complained, and Zuckerberg quickly shuttered the site. He wrote to apologize to two campus women's groups, Fuerza Latina and the Association of Black Harvard Women, saying he didn't anticipate how quickly interest would spread and that he hadn't really thought the idea through. "I definitely see how my intentions could be seen in the wrong light," Zuckerberg wrote in his apology. In a student newspaper article on the controversy, he portrayed himself as little more than a junior computer scientist: "I'm a programmer and I'm interested in the algorithms and math behind it."[13]

The fallout from Facemash was intense, and Zuckerberg was brought before Harvard's disciplinary board, providing yet another parallel with Gates, who faced the same board back in the 1970s over his use of a university computer for company business. Zuckerberg says he and his friends were convinced he would be expelled. Not only could the project be seen as bullying in its premise, but Zuckerberg had trampled over students' privacy by copying and then publishing their photos without permission. The Harvard College Administration Board found Zuckerberg guilty of "improper social behavior" and placed him on probation.[14] One good thing to come from the whole experience, he says looking back, was that he met his future wife, Priscilla Chan, at what was supposed to be his farewell party.[15]

Yet there were other benefits, too, important lessons about building online networks. First, he saw how being part of a relatively small community like Harvard made a social Web site take off—even something as tangential as a hot-or-not site blew up because of the likelihood that a user knew at least one of the participants. Clearly there was an urgent need (call it prurient, call it social, call it mean, call it supportive) to engage online with the people you were living next to. Second, he now grasped what would be the primary challenge in running a successful social site: enticing people to share their personal information online. Hacking that information wasn't a practical solution, of course, but neither was limiting oneself to what was already available.

Zuckerberg took heart from an editorial in the student newspaper, the *Harvard Crimson*, which accepted Facemash for the mischievous idea that it was and just wished that he had limited himself to students who agreed to participate. "Such a site," the *Crimson* wrote, "would have brought joy to attention-seekers and voyeurs alike."[16] Zuckerberg said at one point that this editorial gave him the idea of building Facebook with voluntary participants and built-in privacy restrictions. But within that praise from his fellow students was that fundamental misperception—Zuckerberg's next project couldn't simply appeal to attention-seekers and voyeurs. It had to appeal to everyone or, alternatively, bring out the attention seeker/voyeur in all of us.

During what turned out to be his two years at Harvard, Zuckerberg seemed to be pursuing a curriculum designed for operating a technical/social project like Facebook. He planned to major in psychology—his mother is a licensed psychiatrist[17]—where he could study what made people tick, while taking the most classes in computer science to learn what made machines tick. There would be a dash of Roman history thrown in as well. Not surprisingly, Zuckerberg homed in on the central question in computer science and psychology—namely, can a computer be programmed to think like a human? And if not, why not? His reading in psychology, he said, taught him to be wary of the idea. "The biggest thing that I took away from the psychology classes that I took were how little we know about how the human brain works," Zuckerberg says. "I think that our understanding of the brain is kind of like if you opened up a computer and were like, 'Oh, when you're typing this command this part gets warm.'"[18]

Zuckerberg's views on the potential for artificial intelligence typified the shift in computer science over its fifty-year history. What began as an esoteric discipline largely fueled by the ambitious search for the essence of thought and intelligence, today is a highly practical subject pursued at some level by nearly all undergraduates at a school like

Stanford.[19] Computers haven't proved as profound as John McCarthy and others hoped, but they are certainly quite capable. Computer science has become content to design computers that simulate people well enough that people will acknowledge them, play with them, take their suggestions. In either approach, computer scientists relied on psychology: the early researchers studied the human brain to build thinking machines that worked along the same lines, as hubristic as that may have sounded; the Zuckerbergian social-network operators studied the brain to understand why people do what they do, the better to influence them. As Joseph Weizenbaum's experiment with the computer therapist Eliza showed back in the 1960s, people didn't need much cajoling to open up about themselves to computers—it seemed to come naturally. Weizenbaum was horrified at this misplaced trust by the public and ran away from AI, while Zuckerberg has tried to capitalize on it at Facebook.

Zuckerberg's primary lesson from psychology classes is that people need people and are driven to satisfy this need. We notice and appreciate minute differences in each other's faces and recognize that slight shifts in expression can convey a profound change in mood. "I think that that's something that we often overlook in designing products," he said. "And that's one of the things that I'm just really interested in—and with Facebook—is [that] people are still interesting to other people." People generally want to be with other people and do what other people, particularly their friends, are doing—the network effect, in other words. This interest in both hacking and psychology has allowed Zuckerberg to recognize that a social Web site could become the mechanism for connecting the world, as opposed to, say, a site about money or shopping, entertainment or information: "If you can build a product where people can go and learn about the people around them and share information and stay connected with people then that's something that's super important to people."[20]

However, the Facemash experience chastened Zuckerberg, at least momentarily. He was very nearly kicked out of school in disgrace. He had some serious explaining to do to his parents. And for what? Just to build something? Anything? Yet all the controversy served as great advertising for his programming skills and skills at promotion. In the immediate aftermath of Facemash, a pair of entrepreneurially minded

Harvard undergraduates, the twins Cameron and Tyler Winklevoss, and a friend who had recently graduated, Divya Narendra, sought Zuckerberg's help in completing a student social-networking site they planned to call HarvardConnection and then rename ConnectU once it invariably spread to other schools.[21] None of these three was a hacker, however, a fact that became apparent once Zuckerberg discovered that they had spent almost a year on their idea and had nothing to show for it. Over that same period, Zuckerberg must have created and ditched a dozen projects, including a computer application that would synch up music players so he could try to persuade everyone at Harvard to play the same song at the same time. Why? Because it would be really funny.[22]

This restless energy was what the HarvardConnection team hoped to enlist from Zuckerberg, and they were prepared to name him a partner in the business and give him a share of the company in return for getting the Web site working. But, in truth, no offer from the Harvard-Connection team, other than complete surrender, could fix what was a profoundly unequal relationship. Zuckerberg, the programmer, held all the cards. One day, he emailed to say that he had completed most of the coding and that the site could launch soon, and then he ghosted the team for the next two months. When they managed to get a reply, Narendra says, Zuckerberg would invariably write back to ask for a little more time. After this sputtering couple of months, Zuckerberg broke off the arrangement entirely and informed the HarvardConnection team that he was working on a different project. On January 11, 2004, Zuckerberg registered the domain name, thefacebook.com, his next social-networking Web site for the Harvard campus. On February 4, the site went live, though the Winklevosses learned about it only from an article in the *Crimson* almost a week later.[23] Their first thought, "Well, that sounds like our idea."[24]

The drama that followed from this college breakup has been described in magazine articles, books, and even a movie, *The Social Network*. There was a court case, too, during which Zuckerberg had to submit to a long, contentious deposition and texts were made public in which he mocked his three wannabe partners. The case settled in 2008 with a payment of money and Facebook stock worth tens of millions of dollars to the Winklevosses and Narendra.[25] The central question raised by the court case and the media coverage was: did Zuckerberg steal the idea for Facebook or was an idea like Facebook too obvious for any individual or individuals to own?

One of Zuckerberg's computer science professors, Matt Welsh, defended his former student in a blog post by observing, "Ideas are cheap and don't mean squat if you don't know how to execute on them. To have an impact you need both the vision and the technical chops, as well as the tenacity to make something real. Mark was able to do all of those things, and I think he deserves every bit of success that comes his way." The lesson to Welsh, who left academia to become an engineering manager at Google: "Nerds win."[26] Think about it: Zuckerberg needed nothing that the rich and connected Winklevosses could offer, while they needed everything he had to contribute. For better or worse, these young computer experts now had the power to promote their own big ideas . . . not only about computers, but about everything.

As Zuckerberg anticipated, thefacebook.com filled a vital need on campus, providing a directory for all Harvard students and some basic ways to share information like what courses you were taking and your phone number. It was introduced in early February, and by the end of the month three-quarters of all undergraduates had signed up.[27] Zuckerberg couldn't help taking a shot at the IT team at Harvard, whom he considered cautious bureaucrats in need of student-led disruption. "Everyone's been talking a lot about a universal face book within Harvard," he said in the days after launching his site. "I think it's kind of silly that it would take the university a couple of years to get around to it. I can do it better than they can, and I can do it in a week." Never mind that the Harvard team was delayed in releasing its official online facebook in part because it had to remove the kind of security flaws that allowed a hacker like Zuckerberg to create Facemash. From his nineteen-year-old hacker's perspective, it was clowns to the left of me, jokers to the right.[28]

Initially, what proved fascinating to Zuckerberg about thefacebook was that it provided a platform for social networking that was tied to a particular place, a campus, and a particular group of people, the students there. In this way, the Internet was supplementing offline relationships, not replacing them. The goal of thefacebook "wasn't to make an online community, but sort of like a mirror for the real community that existed in real life."[29] At times, 10 or 20 percent of all Harvard students were logged in at the same time. How could you not belong if all your neighbors did? Thefacebook.com ran as a self-sufficient operation almost from the start, relying on the revenue from a few ads on the site to pay the modest $85-a-month server fees.[30]

This brief time of happy camaraderie lasted the better part of a month. Soon Zuckerberg was, as Professor Welsh describes, "swept up by forces that were bigger and more powerful than anyone could have expected when thefacebook was first launched."[31] Take the central question of whether thefacebook should become a full-fledged business. Zuckerberg was confident that it would not. Though Zuckerberg was never a serious computer science researcher like Brin and Page, who possessed an engineer's commitment to running an ethical, efficient search engine, he shared their instinct that vital Web services like online directories were too important to sully with commerce. A few months into the project, Zuckerberg told a student reporter: "I mean, yeah, we can make a bunch of money—that's not the goal . . . I mean, like, anyone from Harvard can get a job and make a bunch of money. Not everyone at Harvard can have a social network. I value that more as a resource more than, like, any money." He said he had rejected a number of early money-making ideas for his project, like selling email addresses or allowing students to upload résumés and charging companies to search through them. Zuckerberg reassured the student reporter that he would be all right: "I assume eventually I'll make something that is profitable."[32]

Yet Zuckerberg was almost immediately forced to consider whether thefacebook had to grow. For example, what should happen to Harvard students when they graduate? Should they be abandoned? What about students at other colleges who were clamoring for a similar service on their campus? And what about friends on different campuses who wanted to keep up with each other? How about expanding internationally? When thefacebook on March 1 decided to move beyond the Harvard campus, it had a unique advantage at its disposal: status envy. Which schools would be deemed worthy to follow Harvard? As it happens, the next schools thefacebook chose were also elite—Yale, Columbia, Stanford—which stoked a frenzy at the 99.9 percent of American schools that were still excluded. This was a nice twist on the network effect: not only did people want to join thefacebook because others had already joined, but they wanted to join because they suspected they had been deemed unworthy of joining. Studio 54 meets the viral Web.[33]

Growth soon turned into a necessity for thefacebook, however, and the velvet rope was replaced by an open door. In the better part of a year, thefacebook was on nearly every campus in America, and dozens in Europe as well. Turns out that the network effect could be as much of a threat as a benefit. If thefacebook didn't keep growing, another service presumably would spread across the country and what would become of this little, Harvard-only project? Even loyal users would leave in order to have access to all their friends at other schools. Standing still simply was not an option. In fact, Zuckerberg says that the schools where Facebook first expanded may have been elite, but they were in fact chosen because his small team had heard that they were closest to having their own student-run Facebook-type services that could potentially leap to other schools.[34]

Growth and fighting off competitors became the order of the day. Zuckerberg recalled that about a year into the project a rival emerged that called itself College Facebook, which appeared to be a clone of thefacebook, right down to the name. Its plan, he said, was to sign up schools on the West Coast where Facebook hadn't arrived. The response from Zuckerberg's team was to immediately engage the battle under a practice called "lockdown," which meant "we literally did not leave the house until we had addressed the problem," Zuckerberg said. He added that "now it's a little looser of an interpretation inside the company. We don't literally lock everyone inside the office but about as close to that as we can legally get."[35] Once Zuckerberg changed his view on expanding thefacebook to other campuses, rampant growth became the highest priority. His identity as a genius would now be linked to the vibrancy of his project.

Unlike the Google guys, however, Zuckerberg was under little pressure to turn his baby into a business. Even as it grew, thefacebook was able to keep the servers running with little difficulty. And because he was at Harvard, not Stanford, there was no gateway professor to make an introduction leading to an investment offer that he couldn't refuse. The move from cool project to lucrative business seemed to run in relative slow-motion for thefacebook: after his sophomore year, Zuckerberg relocated for the summer to Silicon Valley—he's not exactly sure why, except that it somehow made sense: "Originally when we went out there, we weren't expecting to move out there, we wanted to go out there for the summer because we had this feeling like, 'Okay, all these great companies come from Silicon Valley. Wouldn't it be cool to spend the summer out there and get that experience?'"[36]

Silicon Valley worked its magic, however. At the end of the summer and as successive new semesters approached, the team considered returning to Harvard and then thought: Why not stick it out a little longer? By December 2004, around the time to consider returning for spring semester, thefacebook had a million registered users. Also, the company had already lined up a $500,000 loan from Peter Thiel, after Thiel's friend Reid Hoffman passed on the opportunity of being the first outside investor because of his own social network LinkedIn. Thiel's loan would convert into a 10 percent share of the company assuming thefacebook had grown to 1.5 million users by the end of 2004. It didn't quite reach that goal, but Thiel still decided to convert his loan into an investment.[37] (Hoffman didn't entirely abstain, either. He invested $40,000 for a share of the company that today would be worth hundreds of millions of dollars.)

For someone of Zuckerberg's generation, Thiel was a much more attractive investor than a traditional VC. He had already run a successful company and was outspoken in his belief that a founder should be entrusted with guiding the company he created. Yet for all his faith in Zuckerberg's ability, Thiel also splashed some cold water. He wanted to shake Zuckerberg out of his lackadaisical approach toward running a business. A couple of times in his career as CEO, Zuckerberg would need to have market realities made clear to him. First up was Thiel, who immediately put the twenty-year-old Zuckerberg back on his heels by pointing out that he had mangled how shares were allocated to his partners. "I really knew so little at the time," Zuckerberg recalled. "I mean, when Peter Thiel came in to invest, one thing that he demanded was that all of the founders be on vesting schedules. And I didn't even know what a vesting schedule was. I'd never heard of that."[38] Thiel did leaven his criticisms with rewards, like his gift of an Infiniti FX35. The luxury car could be seen parked outside Zuckerberg's modest apartment, which had a living room with a mattress on the floor and a shower without a curtain.[39]

Facebook became the ideal candidate to test Thiel's theories about the network effect and monopoly power. Even in 2005, Zuckerberg imagined that he was creating a service that the public needed to access every day, unthinkingly, like water or electricity. "We're not trying to create something that people use for a specific purpose," Zuckerberg said at a talk at Stanford barely a year in. "This is a utility that people can use to just find relevant information socially to them."[40] In other words, a social good. Thiel saw a financial good, however, a potential monopoly that could sit above the hypercompetitive marketplace where no one makes any profit.

Thiel's persistent message to Zuckerberg was to grow and grow fast, the equivalent of a traditional utility rushing to lay pipe or cables in order to expand its customer base. "He was good at saying, 'Here's the one thing that matters,'" Zuckerberg recalled about Thiel, who remains on Facebook's board. And what was that one thing? "Connecting everyone as quickly as possible, because network effects were a massively important part of this."[41] During this growth spurt, however, Zuckerberg's project went from a self-supporting site to a project that required large outside investments, the first of which came from Accel Partners in April 2005—more than $12 million for a 15 percent stake, supplemented by $1 million from the partner at the firm, Jim Breyer.[42]

Facebook would need the money to add new programmers and equipment as it kept scaling. Zuckerberg explained his hiring practices at the time, which were exactly what you would expect from a young, privileged hacker and typical for Silicon Valley start-ups. The first quality he sought was "just raw intelligence," that elusive big brain who can build things. Youth, too. Young big brains. "You can hire someone who's a software engineer and has been doing it for 10 years, and if they're doing it for 10 years, then that's probably what they're doing for their life. And I mean, that's cool . . . but if you find someone whose raw intelligence exceeds theirs, but has 10 less years of experience, then they can probably adapt and learn way quickly and within a very short amount of time be able to do a lot of things that that person may never be able to do."[43] The second quality, he said, was devotion to the company's cause so that they will be willing to work really hard.

Here was the typical staffing method of a Silicon Valley start-up that cared not a whit for diversity, racial or gender, or an outside life, and wasn't afraid to tar someone as young as thirty-two as washed up. It's a sobering description of how to staff for a small company, focused on this prized if elusive quality "raw intelligence," which emerges from test scores or an intense interview. But that approach becomes downright scary when you consider that this company now has thousands of employees. There was an awkward scene, too, when a *Rolling Stone* reporter was visiting Facebook's headquarters early on and heard Dustin Moskovitz, Zuckerberg's roommate and cofounder of Facebook, joke about trying to use their Facebook powers to meet attractive women back at Harvard. "Dude, we just got out of a sexual harassment seminar," Zuckerberg snapped at him. These were kids who were still college age, grappling with adult responsibility. The disconcerting fact, however, was that the grownup in charge was Peter Thiel.[44]

Zuckerberg enthusiastically adopted Thiel's advice of pursuing runaway growth, which meant coming to terms with a profound shift in the purpose of his social network. Rather than running a service centered on a particular location, Facebook would increasingly be based on relationships, whether formed online or off. Back in 2005, he expressed his doubts about spreading thefacebook too far, too fast and made a case that a social network should focus intensely on a particular community. "There is a level of service that we can provide when we were just at Harvard that we can't provide for all of the colleges," he said, "and there is a level of services when we're a college network that we wouldn't be able to provide if we went to other types of things."[45]

Once Facebook made this shift from colleges to high schools to particular companies to the wider world, it sought ways to re-create the feelings of connection that came from living close by. The immense amounts of data Facebook collected provided some clues. There was the "ten friend rule," for example; it revealed itself as a pattern among new users as they explored the site, particularly the main newsfeed containing updates about Facebook friends' lives. "Once you had 10 friends," Zuckerberg observed, "you had enough content in your newsfeed that there would just be stuff on a good enough interval where it would be worth coming back to the site. . . . We re-engineered the whole flow of having someone sign up to get all this extraneous shit out of the way and just make it so that the only focus at the beginning is helping people find people to connect with. And we honed what the tools were to get them to do that."[46]

For a puzzle solver and amateur psychologist like Zuckerberg, the access to so much information about people was thrilling. The company put this data to other uses as well. Like PayPal under Max Levchin's guidance, Facebook designed algorithms to determine if a registered user was a real person or a bot accessing the site as part of some scam. "We actually compute how, like a percentage of realness that a person is, and if they fall below a threshold then they're gone," Zuckerberg said in 2005, describing how the algorithms treated humanness fluidly. "It's actually pretty funny. This is something that my friends and I like to do. We just go through and like see how real certain people are who we know are actually real people. We're like, you're only 75 percent real."[47] Another purpose for its algorithms was more akin to what Google does; that is, to help users sift through piles of information to get to what is most relevant. For example, Facebook's algorithms had to determine who among three hundred friends should have his photos shoot to the top of a newsfeed, and whose should largely be ignored.

This Google-like question—what are the most relevant photo results for a particular Facebook user?—points out the similarities between the missions of the two companies. Both are in the information-organizing business. The information that Google organizes is everything that appears on the Web, as well as the material on its own sites, while the information that Facebook organizes is whatever people are willing to share about themselves on its site. Google's initial hurdle was to locate, copy, and store immense amounts of data, using computers that "crawl" the Web; Facebook's hurdle is to persuade the public to be more forthcoming. "You can't just send a Web crawler around and learn what's going on with people," Zuckerberg said, "you have to build tools that give people the power to share that content themselves."[48]

In July 2006 Yahoo offered a cool billion for Facebook, which Thiel pushed Zuckerberg to at least consider. Marc Andreessen says he was a rare voice encouraging Zuckerberg to follow his instincts and keep trying to expand Facebook. "The psychological pressure they put on this 22-year-old was intense," said Andreessen. "Mark and I really bonded in that period, because I told him, 'Don't sell, don't sell, don't sell!'"[49] When Zuckerberg walked into the board meeting that considered the Yahoo offer, Thiel recalled, he said, "Okay, guys, this is just a formality, it shouldn't take more than 10 minutes. We're obviously not going to sell here."[50]

Looking back, Zuckerberg says he regrets that he even entertained the idea of selling. "I mean if you don't want to sell your company, don't get into a process where you're talking to people about selling your company," he said, laughing. But the entire experience, he says, gave him a way to assess the core values of his team, and then shed those who tipped their hand that they were motivated exclusively, or primarily, by getting a share of a billion-dollar fortune. "That's an awesome outcome, but that's not what I was in it for and I wanted people around me for whom that's what they were in it for was to build a company for the long term," he said.[51] Thiel wasn't tarred by his association with the Yahoo offer. He remains a key advisor to Zuckerberg, though Zuckerberg clearly enjoys needling him about what a mistake selling Facebook would have been, even worse than selling PayPal for $1.5 billion.[52] On the other hand, Andreessen was invited to join Thiel on the Facebook board two years later, in 2008.

The Yahoo offer was so scarring because it confirmed what Zuckerberg distrusts about the predominant view in Silicon Valley—so many there appear to be in it for the money! "If I were starting now, I just would have stayed in Boston I think," he told an interviewer in 2011, when he was still in his mid-twenties. Silicon Valley clearly had the experienced hands to help a nineteen-year-old scale his college project—the accountants, the lawyers, the data centers, the programmers, the steely minded VCs. But, he adds, "there's aspects of the culture out here where I think it still is a little bit short term focused in a way that bothers me. You know, whether it's like people who want to start companies . . . not knowing what they like, I don't know, to like flip it."[53] Zuckerberg is clearly conflicted by the idea of running a business. A year earlier, in 2010, he defended the role of the market in Facebook's growth, speaking in engineering terms. "Building a company," he said, "is one of the most efficient ways in the world that you can kind of align the incentives of a lot of smart people, towards making a change."[54]

In May 2012, these abstract considerations suddenly became real. Facebook stumbled through its IPO, and the scenario was similar to what Google's founders were put through when they acted as though they were running a computer lab rather than a business with investors who were only interested in profits. Their resistance to carrying advertising on their beloved search engine, or to accepting an outside CEO, melted away. Likewise, when Facebook was failing to deliver the expected level of profits after the stock offering, Zuckerberg was forced to adjust his attitude. He assigned a top engineer to the task of giving a boost to advertising, just as the Google guys had. "Wouldn't it be fun to build a billion-dollar business in six months?" Zuckerberg asked the engineer, Andrew Bosworth.

Like Brin and Page, Zuckerberg made a big concession on advertisements, reversing his view that they should not be permitted in the main newsfeed unless a friend had "liked" the product. The newsfeed had been sacrosanct—a process that had its own logic, flowing from the interests of your particular set of friends, independent of business considerations—much the way Brin and Page had treated search results. Interestingly, when Zuckerberg bent on this principle, he, like Brin and Page, later cited research showing that users were actually happier with the new ads.[55]

These detours into business calculations must have struck a blow to the egos of Know-It-Alls like Zuckerberg and Brin and Page. From the start, they have been the indispensable ones; they called all the shots. Yet when the market turned, a hard truth emerged: in fact, others were calling the shots. You could take the long view and see this as nothing more than the Tesla lesson, which taught that even the most free-thinking innovators had to spend some serious brainpower pleasing investors lest they lose the freedom to work on the really important questions. In Google's case, those questions involve applying artificial intelligence in ever more elaborate ways, which was what brought them to the Stanford computer science department in the first place.[56] Zuckerberg seems to have even more ambitious goals that he wants to be free to pursue: connecting the world "so you have access to every idea, person and opportunity."[57]

In the immediate years that followed, Zuckerberg returned to the goal of global connection and communication, something his company pursues through technical improvements, like the algorithms designed to entice users to share more data or the drones and satellites that provide rudimentary Internet access to remote areas so that new users can share personal information about themselves. At times, too, Facebook has relaxed the privacy settings for the information stored in its computers, confident that once more material is being shared, its users will be pleased and accept the new standards. Users have at times objected, however, and then Facebook apologizes and returns to some version of the old settings.[58] But it keeps trying, confident that, as the ten-friend rule suggests, once people experience the benefits of sharing information they will want to share more, too. In fact, Zuckerberg has observed that the amount of information being shared online has doubled every year—a discovery his friends have called Zuckerberg's law, in a tip of the hat to Moore's law, which predicted the steady increase in microchip processing speed.

This new law, Zuckerberg believes, has a disruptive potential similar to that of Moore's law. As we share more information via digital tools like Facebook, we as a species are enhancing our ability to connect and feel empathy for others. If a computer is about extending a person's ability to think—providing what Steve Jobs called "a bicycle for our minds"[59]—a social network "extends people's very real social capacity," Zuckerberg argues.[60] As evidence, he points to a change Facebook has detected in the so-called Dunbar number, which was originally proposed to suggest that there is a cap to the number of close friends a person can have at one time. Based on observations of other primates and correlating brain size to the number of members in a species' social group, the anthropologist Robin Dunbar argued that humans were capable of maintaining genuine, empathetic relationships with up to about 150 people.[61]

Facebook, Zuckerberg contends, has raised that number because of the ease and efficiency in making and keeping online friendships. "Naturally when people sign up the average amount of friends that they get is around 150," he said, "but then over time it can expand and you can keep in touch and stay in touch with many more people."[62] This is an oddly quantitative way of suggesting that there has been an "improvement" in the human ability to feel empathy. To start, this presumes that maintaining many good friendships is superior to nurturing just a few. But this bias shouldn't necessarily be surprising. Silicon Valley is a culture that evaluates intelligence with a number and business success with a number and attractiveness with a number. Why not sociability, even if it means draining friendship of all that is unquantifiable—the depths of shared history, a mysterious sympatico?

In a letter Zuckerberg and his wife wrote to their daughter, Max, they explain how expanded Internet access will help the public in so many ways, extending life spans, lifting hundreds of millions out of poverty, broadening educational opportunities. They are expressing a generous impulse, but it necessarily comes in a disturbing package: a world filled with isolated individuals—geographically and emotionally—who are expected to fend for themselves rather than look to the community for sustenance. In the Zuckerbergian vision, an Internet connection becomes a lifeline: "It provides education if you don't live near a good school. It provides health information on how to avoid diseases or raise healthy children if you don't live near a doctor. It provides financial services if you don't live near a bank. It provides access to jobs and opportunities if you don't live in a good economy. The Internet is so important that for every 10 people who gain Internet access, about one person is lifted out of poverty and about one new job is created."[63]

This seems a strange way to think of helping a person, as opposed to helping, say, a computer. For better health care, we must build clinics and hospitals and train doctors, as well as improve access to self-treatment via the Internet. For a better education, we must build schools and hire more teachers and train and compensate them better, as well as introduce self-teaching via the Internet. For a better work experience, one that ensures that workers are treated fairly and given responsibility as part of a motivated team, we must promote unions and mutual support in the workplace rather than encourage an individual to become a "start-up of you" forced to sell her services online at the whim of the market. Any path to health, education, and wealth must include neighbors, the people who are part of an actual community, not just a virtual one. Even the great Zuckerberg needed a personal tutor at first to learn how to program.

Education has been a particular focus of Zuckerberg's philanthropy, and he has brought his familiarly individualistic, mechanistic view of how people behave to both teachers and students. There was his well-publicized foray as a donor to the Newark, New Jersey, school system, when he tried to tie his $100 million gift to a series of policies toward teachers, like awarding bonuses of up to 50 percent of salary to teachers based on performance. An educational algorithm, if you will. He also proposed eliminating certain other policies—algorithms that he didn't like—like teacher seniority, which is an automated way of saying that age and experience should entitle you to advantages.

Zuckerberg's positive program of using bonuses to extract better performances from teachers, and encouraging more ambitious people to become teachers, faced a few problems and was never adopted by the school system. First, the district would be unable to sustain such a program once Zuckerberg's donation ran out—it was too expensive, when applied to a unionized school system. Furthermore, there was no precise way to measure the quality of a teacher; test results can be gamed and can reflect the preparation and home life of the students rather than the performance of the teacher. Finally, bonuses were not likely to be as effective as Zuckerberg imagined in motivating a group like teachers, as opposed to members of a Silicon Valley start-up. Teachers explained that the money wasn't such a direct influence—it wasn't that they didn't want or believe they deserved more pay, it was just that bonuses wouldn't lead to "better teaching" perforce, as Zuckerberg's view of human motivation would expect. Having a supportive principal was the greater motivation for a teacher to work hard than money.[64]

Zuckerberg viewed students, not only teachers, as individuals who approached education in isolation. "You'll have technology that understands how you learn best and where you need to focus," Zuckerberg and Chan write in their letter to their daughter about how education should improve. "You'll advance quickly in subjects that interest you most, and get as much help as you need in your most challenging areas. You'll explore topics that aren't even offered in schools today. Your teachers will also have better tools and data to help you achieve your goals." What Zuckerberg and Chan call personalized learning—guided by algorithms applied to a student's work—"can be one scalable way to give all children a better education and more equal opportunity."[65] When the term *scalable*, which defines something unbound by all-too-human frailties, is applied to a something as human as preparing a child for life, perhaps a warning flag should go up.

Interestingly, however, when Zuckerberg and Chan discuss a school they are planning in East Palo Alto, California, after the debacle in Newark, they include working with health centers, parent groups, and local governments to ensure that "all children are well fed and cared for starting young." In order to improve the quality of education, they write, "We must first build inclusive and healthy communities."[66] A sensible approach, no doubt, but also one that won't scale and is more traditional than disruptive. In fact, traditionally, we would expect this kind of work to be carried out by local, state, and federal governments rather than concerned billionaires dipping into their monopoly profits.

Assistance for the developing world from Zuckerberg is still stuck in the earlier, algorithmic, phase. For example, he expresses mathematical certainty in the benefits from Facebook's Free Basics, which offers a free, stripped-down version of the Internet, including Wikipedia and Facebook, to those who can't afford any access to the Internet. Zuckerberg again cited his one in ten formula for poverty reduction from Internet access: "There are more than 4 billion people who need to be connected and if we can connect them, then we'll raise hundreds of millions of people out of poverty," he said in defense of Free Basics.[67] The Indian government nonetheless rejected the idea, in part motivated by more than 750,000 emails complaining that the service would create a poor Internet for poor people.[68] Critics in India, like one group called Savetheinternet.in, saw a more nefarious purpose behind Facebook's offer, which it described as "Zuckerberg's ambitious project to confuse hundreds of millions of emerging market users into thinking that Facebook and the Internet are one and the same." All children, even poor children in India, the group wrote, "deserve the same experience and opportunities when it comes to an open and free Internet, as much as their urban or richer peers."[69]

Facebook board member Marc Andreessen was enraged and shared his feelings, via Twitter. The Indian government clearly didn't know what was best for its people. "Another in a long line of economically suicidal decisions made by the Indian government against its own citizens," read one Andreessen tweet. Another, which drew the most complaints, "Anti-colonialism has been economically catastrophic for the Indian people for decades. Why stop now?"[70] Like Thiel, another Facebook board member, was known to do, Andreessen was merely stating what everyone around him thought, though he later apologized to the nation of India and the Indian people. Because Andreessen's medium was Twitter, he couldn't say he was misquoted or misunderstood by an interviewer; rather he insisted that he was "100% opposed to colonialism, and 100% in favor of independence and freedom, in every country, including India."

Zuckerberg quickly condemned Andreessen's comments and stated his objection to the Indian government's decision on Free Basics with much more finesse. He wasn't criticizing the Indian people, but, on the contrary, was speaking up for those without the power to speak for themselves. "Remember," he said, "that the people this affects most, the four billion unconnected, have no voice on the Internet. They can't argue their side in the comments below or sign a petition for what they believe. So we decide our character in how we look out for them."[71] Zuckerberg, along with his wife, makes a similar pledge to his daughter: "We must engage directly with the people we serve. We can't empower people if we don't understand the needs and desires of their communities."[72]

Of course, another way to empower people is simply to give them power. Power to use or even misuse. But Zuckerberg wasn't speaking as a politician. Or even as an entrepreneur. His vision of an interconnected humanity, whose members share details of their lives within the Facebook platform, wasn't something that emerged organically from the people, or even with some gentle nudging from the people's representatives, the government. Zuckerberg was conceiving a new online civilization before our eyes and, if he succeeded, he would be responsible for something grander than Julius Caesar or even Bill Gates could ever have imagined.

THE FUTURE

"Local, small-scale, active"

I n college, I used to marvel at American history books that explained panics like the Salem witch trials or religious revivals like the Second Great Awakening by pointing to social and economic shifts that were profound but largely undetected by the people living through them: new shipping routes linking up to the Caribbean, say, or the settlement of the West. What must it be like to be buffeted by such forces and be none the wiser? Poor fools. Yet the past twenty-five years of radical social, political, and economic change since Tim Berners-Lee introduced the Web have given me new sympathy for such pawns of fate. Our behavior has changed, too. Extremist mass movements and social panic are gathering strength. People are distrustful and angry, they feel isolated and vulnerable to exploitation. The Web itself has encouraged this troubling state of affairs, yet we rarely stop to question how and why the Web developed the way it has. If *The Know-It-Alls* can achieve one thing, it will be to demystify the origins of the harsh market-based values being pumped out by Silicon Valley, which can seem irresistible. They are not. At crucial moments, a relatively few well-positioned people made decisions to steer the Web in its current individualistic, centralized, commercialized direction, rejoicing at the profound disruptions this has produced along the way.

Berners-Lee intended the Web to be something like the opposite—decentralized, collaborative, and minimally commercial. He assumed a generosity of spirit that would drive people to work together to help their neighbors; in this way, the Web would add to the world's all-too-thin connective tissue. And he was not alone in those values. Google's founders, Sergey Brin and Larry Page, saw the Web as having the potential to revolutionize the world's access to knowledge and warned in biting language about the harm that would come from the commercialization of search engines. At the start, they viewed market-based competition as the enemy of their chosen task of organizing the world's information, not its friend. Likewise, Mark Zuckerberg originally saw Facebook as being most effective as a local tool that supplemented real-world relationships, the way it worked when it was still called thefacebook.com and only operated on university campuses.

The founders of Google and Facebook were seduced by the power and money offered by venture capitalists and agreed to reorient their socially minded designs to become profit generators. Turns out that idealistic projects from ambitious, talented young people can become quite lucrative after their leaders have ditched the idealism. The tracking of users' Web searches, which Google initially deployed in the name of "perfect" search results, became tools for placing the most effective advertising. Facebook's fascination with its users' interests and preferences, originally intended to forge friendships, also became grist for highly targeted advertising. Other, less-idealistic entrepreneurs, like Marc Andreessen, Peter Thiel, and Jeff Bezos, managed to inject just enough tech idealism into their profit-making schemes—about empowering consumers, for example—to create lucrative companies of their own. Together, these Know-It-Alls represent a sixty-year climb-down in the goals for computer science and artificial intelligence, from teaching machines to think like humans to teaching them how to help some humans make money by imitating how other humans think.

Thiel likes to describe those diminished ambitions in today's Silicon Valley by saying, "We wanted flying cars, instead we got 140 characters."[1] Well, Thiel may want flying cars, but computer pioneers like John McCarthy wanted something much more extravagant, a new form of intelligence that could make its way through the world. This was a goal worth pursuing not only because artificially intelligent machines would be so cool—more cool, even, than flying cars—but because they would be so profound. Just think of the things these thinking machines could think. But you can't. They've never been thought before. McCarthy and his peers weren't focusing on how to profit personally, but on how society would benefit from having a supreme intelligence to guide it.

In a lecture back in the 1960s, McCarthy made clear that he wasn't worried about a time when computers would be smarter than people, as many tech leaders are today. So what does he suggest we do during our first encounter with these superior machines? Well, obviously, we should ask the machine itself for advice: We created you, oh wise machine, now how should we get along? Armed with the answer, whatever it may be, "we could act accordingly and with greater intelligence than now."[2] In this ultimate sense, McCarthy and his idealistic hackers were laying the groundwork for intelligence to be the most important quality in a leader. He wasn't building an empathy machine that "felt" more intensely than humans. But McCarthy at least presumes a gentleness within these great new intelligences. Of course they will want to help us with our troubles.

By contrast, here is how Peter Thiel reacted to the prospect that genuine artificial intelligence would arrive one day. "The development of AI would be as momentous as the landing of extraterrestrials on this planet," Thiel told an audience on Reddit, shuddering at the thought. "If aliens landed, the first question would not be about the economy!" Presumably, the first question would be about our safety, as in, "Will AI be friendly?"[3] If you see great intelligence, at its heart, as about the power to control and dominate others, you understandably would tremble at encountering a bigger brain. Suddenly the tables would be turned. This reflects the Know-It-All playbook, which closely resembles the Donald Trump playbook, with its emphasis on domination and disruption and life as a competition. McCarthy and the early AI pioneers conceived of their research as part of a collective human quest—funded by the government, destined to help all of us. The current crop of tech leaders, however, see these civilization-altering quests as personal. Consider how the billionaires Elon Musk and Jeff Bezos are taking on the ultimate collective human quest, space exploration, through private companies, SpaceX and Blue Origin.

McCarthy isn't blameless, of course, when it comes to the online culture we all endure these days. Hiding behind their reverence for supposed "raw intelligence," he and his hacker acolytes created a computer lab ethos that was unwelcoming to people who were different, particularly women and minorities. As we've seen, hacker-led tech companies re-created those patterns. McCarthy and the hackers were fierce toward those who couldn't cut it, and today the Web is overrun with aggressive, nasty comments, which are tolerated by large companies like Twitter as being part of a robust debate. Furthermore, by challenging the very idea of authority, McCarthy and the hackers gave comfort to the most destructive impulses of start-up founders.

But the early hackers' indifference to wealth and personal power meant that the harm from these toxic ideas was largely contained within the walls of the computer lab. In the 1970s, McCarthy's artificial intelligence lab was five miles off campus, its members' heads even higher in the clouds. "In those old glory days at SAIL," wrote the cognitive scientist Douglas Hofstadter, "speculations about the undecidability of Fermat's Last Theorem mingled freely with pictures of Cantor sets, Cantor dusts, Sierpinski gaskets, snowflakes, flowsnakes, blinkers, traffic lights, gliders, and glider guns, while somewhere down the hall, Principia Mathematica, Gödelian incompleteness, the halting problem, and other Ouroborous-like diagonalizations nestled up to Stanislaw Lem's 'Cyberiad,' Newcomb's problem, teleportation fantasies, and free-will puzzles, and elsewhere, Necker Cubes, Soma cubes, Go boards, impossible objects."[4] McCarthy's lab knew its place—it was part of a system of open research conducted at university labs with the support of government grants. Even the great anti-censorship fight at Stanford championed by McCarthy was an attempt to protect the early Internet, which he considered home turf. McCarthy wanted the lab's ideas tested, challenged, and ultimately used to help the public; he didn't will his or someone else's ideas to run the world single-mindedly. And that, in short, is what changed.

Who would have thought to impose the values of the lab on the world, anyhow? That person was Frederick Terman, Stanford's ambitious provost who didn't understand why his university and its students and faculty shouldn't profit from their brilliant ideas. He promoted the idea that engineers should become rich and powerful, starting with two of his early students, William Hewlett and David Packard. Stanford benefited. Shareholders benefited. Society benefited. If engineers set the agenda for the country, and the world, then at last we would be certain that the brightest would be in charge, as his father, Lewis, the student of "gifted children," had dreamed. Ordinary folks wouldn't need to be weighed down by questions they couldn't understand, anyhow. On efficiency grounds, Terman's vision might make sense. On humanitarian grounds, less so. Do we really want engineers—and their hyperrational comrades—to determine how we live? Democracy, for all its faults, is the best way to ensure that the public is being served by its leaders. No one can look out for your interests as diligently as yourself and no engineer is so unemotional as to be acting purely rationally without bias or self-preservation.

And, let's face it, an un-self-aware engineer is really the best-case scenario. The people who populate *The Know-It-Alls* have dreams of world domination and seem especially poor candidates to set the priorities for society. Consider Peter Thiel, who was once asked if the super-rich were happier than the rest of us. He first answered by saying enormous wealth can cut both ways when it comes to happiness, and then added, "I've always questioned the premise of the question. I'm not sure whether subjective happiness should be the most important metric at which we evaluate things. There's many other metrics we can use." His personal crusade has been to harness science to forestall death. Life span is a favorite metric. Whether Thiel is happy or not with existence, he is certain that he wants more of it. "I actually think life is something that's worthwhile in and of itself," he told an interviewer. "Death is kind of a bad thing, in and of itself, so even if I was adrift and had no sense of what I was doing at all, I would still all else being equal, hopefully prefer to live a lot longer."[5] His vision of the future represents a slight improvement on Hobbes—nasty, brutish, and long.

Then there is Mark Zuckerberg, who in a letter written with his wife, Priscilla Chan, in 2015 to his newborn daughter, Max, laid out what he hoped for her future and our future. His aspirations were oddly quantitative. Max's birth was the occasion for Zuckerberg and Chan to announce that they were creating a private corporation, the Chan-Zuckerberg Initiative, to invest and donate the billions of dollars he had made from Facebook. The purpose of the initiative, they wrote in the letter to Max, was to advance human potential and promote human equality. And, in detailing the kind of work they would be supporting in those causes, Zuckerberg and Chan proposed "pushing the boundaries on how great a human life can be," including the goal of being able to "learn and experience 100 times more than we do today."[6]

What a beguiling notion! A hundred times more to experience and learn than is typical today—like suddenly living your life in high-def. Facebook is helping to bring about that future as best it can. Already, people have more close friends than once was thought possible, according to Zuckerberg, and those friends are sharing more about themselves, stories, photographs, new hobbies, new relationships. You do the math. Based on your access to Facebook, how many more experiences have you already experienced, how many more facts have you learned? With its investment in virtual-reality technology, Facebook is hoping to bring more of the world in front of our eyes and with life-like clarity.

This line of reasoning does inevitably raise a few questions, such as: Are experiences really so fungible? Is watching a hundred sunsets a day through a virtual-reality headset, for example, one hundred times the experience of lingering over the sun as it sets once and only once outside your window? On first blush, this quantitative approach to life is more peculiar than pernicious. Why exactly would someone want a life overstuffed with experience as Zuckerberg with his social networks and virtual-reality devices and Thiel with his life-extension research are promising? What is pernicious, however, is the incidental destruction, the collateral damage, as these leaders pursue their dreams by centralizing our relationships—personal, economic, political. Local stores, local newspapers, local unions, local political organizations are swept away as tech companies centralize profits.

We are lonely and crave connection. We are confused by life and crave meaning. We are anxious and crave the feeling of being cared for. We are angry and crave the ability to trust. Internet giants like Facebook are eager to step in. In a 2017 manifesto, Zuckerberg described the online communities facilitated by his social-networking platform as "a bright spot" in the darkening global landscape. Not only can Facebook "strengthen existing physical communities by helping people come together online as well as offline," he wrote, but "in the same way connecting with friends online strengthens real relationships, developing this infrastructure will strengthen these communities, as well as enable completely new ones to form." Facebook is nothing less than our salvation. Of course, the reason Zuckerberg was writing this particular statement to the world was the criticism of Facebook's role in encouraging extremist, hate-filled political movements during the 2016 presidential election and what he himself described as "surveys showing large percentages of our population lack a sense of hope for the future."[7]

The one connection that Zuckerberg wasn't prepared to make was between the growth of a centralized, commercialized Web and the lack of hope and sense of isolation he sees all around him. But that connection is what the preceding pages have been about. When you have been isolated and treated as a commodity by big companies, whether as a target of advertisers or as a potential "start-up of you"; when you lose the ability to protect your fellow citizens online and off because of the assault on business regulations and a social safety net; when you see obscene wealth delivered to the hands of a few; when hostility and incivility is the coin of the realm in online conversation; when racial, ethnic and gender inclusion are considered assaults on freedom. . . . Well, naturally, you lose hope and feel unprotected.

Where do we go from here? Looking to Facebook and the other tech giants for solutions would be madness. Europe provides a model, with its greater protections for individuals against tech companies' ability to exploit the information it collects. European governments have also tried to prevent Silicon Valley companies from holding monopoly power in the marketplace, in part to support their own start-ups. These are all steps in the right direction.

But rather than offer a set of policy proposals, I would repeat a prescription for a just society that begins with "a commitment to the local, the plural, the small scale and the active."[8] Those are the qualities the Web must have, even if it means cutting off the flow of revenues to giant companies like Google, Facebook, Amazon, and eBay. We can't tolerate an Internet, or a society, led by a few self-proclaimed geniuses claiming to serve mankind. The Internet can and must work for us, instead of the other way around, through a diversity of voices and platforms free to organize and collaborate on their own rather than through a few centralized services. This way the Internet can help build the social connective tissue we so desperately need, as Berners-Lee originally intended.

A NOTE TO
THE READER

Careful readers will notice that in *The Know-It-Alls* I have tried hard not to use gender-neutral pronouns or mix up the "shes" and "hes" in some sort of equalized fashion. Certainly, such an effort would send a message of inclusion, and that is precisely why it would be so inappropriate for a book like this. The story of how Silicon Valley leaders came to their libertarian worldview is the story almost exclusively of men: the hackers who first mastered computers and promoted extreme individualism, as well as the venture capitalists and tech start-up leaders who brought those innovations to market. Fittingly, the political theory they promote is one that necessarily belittles the contribution of women as part of its fantasy of extreme individualism, in which adult men arrive in the world magically with no debt to anyone. We, author and reader alike, must not lose this thread of systematic exclusion even if it is intended to promote a fairer world going forward.

I am not the first to bring this argument to my writing about Silicon Valley. In her review of *The Hard Thing About Hard Things*, a business advice book written by Ben Horowitz, founding partner of the venture capital firm Andreessen Horowitz, Diane Brady took pains to praise Horowitz for his work in support of women's rights, but nonetheless expressed her bafflement at phrases like "tough times separate women from girls" and other conspicuous efforts by Horowitz to treat the sexes the same in his writing. "Horowitz's persistent use of 'she' comes off as social satire in a world with so few actual women," Brady writes in *Bloomberg Businessweek*, a world where "'her' is more likely to refer to your operating system than to your business partner."[1]

Not surprisingly, this call for self-reflection by the men who lead Silicon Valley was met with contempt, most notably by Marc Andreessen, Horowitz's business partner, who wrote with typical subtlety on Twitter: "Mind-bending review of known misogynist pig @bhorowitz's new book :-)".[2] Yet surely there can be a system inherently unfair to women without each of its leaders being "known misogynist pigs." Calling out superficial attempts at equality is a vital first step toward promoting real change in a field where there is so much self-satisfied resistance.

ACKNOWLEDGMENTS

This book grew out of my writing and editing at the *New York Times*, particularly the six years or so when I was responsible for the Link by Link column, which tracked the influences—amusing, wrenching, terrifying—of Internet culture. Bruce Headlam, the media editor at the time, saw that this column would suit my interests and shepherded me and my copy expertly. When I turned my focus to *The Know-It-Alls* I had a number of inspiring discussions: Lee Vinsel was generous with his time as he helped me understand how my book could fit into the larger critique of the cult of innovation; Paulina Borsook walked me through her experiences writing *Cyberselfish*, which was prescient in describing the deficiencies of a society led by Silicon Valley entrepreneurs; Sue Gardner gave me encouragement that I was on the right path based on her own time in the Bay Area; and my cousin Carl Shapiro, who died as I was writing the book, was my special guide through the history of computers and the Internet.

I had the benefit of old and new friends and family members who either read drafts or heard outlines and interrogated my premises: Robert Mackey, Todd Schiff, Martha Bridegam, David Malaxos, Alan Cohen, Lori Cohen, David Crossland, Kevin Kolben, Jenna

Wortham, Liam Wyatt, Joseph Reagle, Sarah Szalavitz, Victoria Baranetsky. Jeff Wise hashed through the ideas in this book week upon week. Eric Kaplan and Joseph Tedeschi were invaluable close readers of the entire manuscript, always encouraging, always questioning. My agent, Rafe Sagalyn, kept this neophyte on the right path.

My former *Times* colleague, John Markoff, kindly shared his recordings of his interview with John McCarthy. Thanks, too, to Michael Zimmer, whose Zuckerberg Files, hosted by the University of Wisconsin, Milwaukee, expertly tracks and transcribes the talks and interviews of Mark Zuckerberg. The *Stanford Daily* should be commended, too, for its excellent digital archive.

I relied on the reporting of many fine technology writers and historians to make my case about the Know-It-Alls, even if these reporters wouldn't necessarily agree with my conclusions. Steven Levy's pathbreaking work, *Hackers*, is the ur-text for understanding the young men who caught the computer programming bug and convinced themselves these machines could save the world. C. Stewart Gillmor's chronicle of Frederick Terman at Stanford manages to be a deeply reported history of a man and a university. In *Once You're Lucky, Twice You're Good*, Sara Lacy created an indelible portrait of the ferment of the so-called Web 2.0 years. George Packer is insightful in explaining how Peter Thiel belongs in the story of Silicon Valley values. My arguments were shaped by the detailed research of, among others, Rebecca Lowen and Paul Edwards, and, more overarchingly, by the intellectual history of Daniel T. Rodgers, whose *Age of Fracture* captures the hyperindividualism of America at the end of the twentieth century even if the words "Internet" and "Web" never appear. The humane values of Joseph Weizenbaum, the world-weary, lapsed artificial intelligence believer, were a touchstone.

At The New Press, my editor, Carl Bromley, was a steady hand and ready guide for me on this project, starting with the inspired title. His colleague, Jed Bickman, was an important early reader of my drafts. Diane Wachtell has long been a friend and advocate. Jessica Yu has been a model in getting the ideas in this book in front of the public, while Emily Albarillo has been a model in getting the words in this book in front of the public.

My father, Stuart Cohen, is my example of being open to what the world has to offer; for as long as I have known him, he unfailingly accentuates the positive. Harlan Cohen and Adam Cohen are both a spur and a safety net, as only older brothers can be. My wife, Aviva Michaelov, is a source of joy, pride, and profound friendship, the pillar in my life supporting so much more than the writing of a book, as all-encompassing as that task may have seemed at times.

During the writing of this book, I lost my mother, Beverly Sher Cohen. Her questioning spirit and joy in life and language surrounded me as I was writing, much the way I see her vividly in the smiles, giggles, and occasional words of reproach from her granddaughters, Kika and Nuli.

NOTES

Introduction

1. "To Serve Man," teleplay by Rod Serling, based on a short story by Damon Knight, *The Twilight Zone*, aired March 2, 1962.

2. "Minimum Viable Product," by John Altschuler, Dave Krinsky, and Mike Judge, *Silicon Valley*, season 1, episode 1, aired April 6, 2014.

3. "Proof of Concept," by Clay Tarver, *Silicon Valley*, season 1, episode 7, aired May 18, 2014.

4. Google Privacy and Terms, "Welcome to the Google Privacy Policy," last modified August 29, 2016: http://www.google.com/policies/privacy.

5. Originally called Internet.org, the project's home page is: https://info.internet.org/en/story/free-basics-from-internet-org.

6. Video, May 4, 2015, available at Mark Zuckerberg's Facebook page: https://vimeo.com/126762664.

7. Adi Narayan, "Andreessen Regrets India Tweets; Zuckerberg Laments Comments," Bloomberg.com, February 10, 2016.

8. Susan Moller Okin, *Justice, Gender and the Family*, New York: Basic Books, 1989, p. 75.

9. Reid Hoffman and Ben Casnocha, *The Start-up of You: Adapt to the Future, Invest in Yourself, and Transform Your Career*," New York: Crown Business, 2012, pp. 159–161.

10. Ibid., pp. 8–9.

11. Tim Berners-Lee with Mark Fischetti, *Weaving the Web: The Original Design and Ultimate Destiny of the World Wide Web*, New York: Harper, 2000.

12. David Streitfeld and Malia Wollan, "Tech Rides Are Focus of Hostility in Bay Area," *The New York Times*, February 1, 2014, p. B1.

13. "Inside the Extravagant Wedding of Sean Parker and Alexandra Lenas," *Vanity Fair*, August 1, 2013.

14. Peter Fimrite, "Vinod Khosla Wants $30 Million for Martins Beach Access," *The San Francisco Chronicle*, February 23, 2016.

15. Katherine Bindley, "David Sacks, Yammer CEO, Hosts Extravagant Birthday Party," *The Huffington Post*, June 19, 2012.

16. Steven Bertoni, "Instagram's Kevin Systrom: The Stanford Billionaire Machine Strikes Again," *Forbes*, August 20, 2012.

17. Rich McCormick, "Ghostbusters Star Leslie Jones Calls for Stronger Twitter Guidelines after Racist Abuse," *The Verge*, July 19, 2016.

18. Peter Thiel, "Address to Republican National Convention," Cleveland, OH, July 21, 2016, C-Span: https://www.c-span.org/video/?c4612796/peter-thiel-addresses-republican-national-convention-proud-gay.

19. David Streitfeld, "Peter Thiel to Donate $1.25 Million in Support of Donald Trump," *The New York Times*, October 15, 2016.

20. Elon Musk, post to Twitter, Jul 31, 2012, @elonmusk.

21. Erik Wemple, "Peter Thiel's Media Critique: Reporters Take Trump's Statements 'Literally' but Not 'Seriously,'" *The Washington Post*, October 31, 2016.

22. See Dylan Matthews, "Jeff Bezos Is Buying the Washington Post. Here's What You Need to Know About the Sale," Wonkblog, *The Washington Post*, August 6, 2013; David A. Graham, "The Politics of New Washington Post Owner Jeff Bezos," *The Atlantic*, August 5, 2013.

23. See Steven Levy, *Hackers: Heroes of the Computer Revolution*, Penguin Books, New York, 2001 (original 1984).

24. Richard W. Lyman, *Stanford in Turmoil: Campus Unrest, 1966–1972*, Stanford, CA.: Stanford University Press, 2009, p. 5.

25. Barbie Fields, "Frederick Terman—A Living Bay Area Legend," *The Stanford Daily*, October 7, 1977, p. 1.

26. Carolyn E. Tajnai, "From the Valley of Heart's Delight to the Silicon Valley: A Study of Stanford University's Role in the Transformation," Stanford University, Department of Computer Science, 1996: http://forum.stanford.edu/carolyn/valley_of_hearts.

27. See C. Stewart Gillmor, *Fred Terman at Stanford: Building a Discipline, a University and Silicon Valley*, Stanford, CA: Stanford University Press, 2004.

28. David Orenstein, "Computer Science@40: Faculty, Alumni Celebrate Life-Changing Advances," *Stanford News*, April 5, 2006.

29. Telephone interview with Richard Weyhrauch, May 23, 2016.

30. See Steven Levy, *In the Plex: How Google Thinks, Works, and Shapes Our Lives*, New York: Simon and Schuster, 2011.

31. Ken Auletta, "Get Rich U.," *The New Yorker*, April 30, 2012.

32. Tonya Garcia, "Amazon Will Account for More Than Half of 2015 E-Commerce Growth, Says Macquarie," *MarketWatch*, December 22, 2015.

33. Max Chafkin, "What Makes Uber Run," *Fast Company*, September, 8, 2015.

34. Ginia Bellafante, "Airbnb's Promise: Every Man and Woman a Hotelier," *The New York Times*, December 17, 2014.

35. Statistics from Facebook's Newsroom: http://newsroom.fb.com /company-info.

36. Stanford University, "James Breyer/Mark Zuckerberg Interview, Oct. 26, 2005, Stanford University" (2005), Zuckerberg Transcripts.

37. Computer History Museum, "The Facebook Effect (interview with Mark Zuckerberg and David Kirkpatrick)" (2010), Zuckerberg Transcripts.

38. Douglas Bowman, "Goodbye, Google," March 20, 2009, at Stopdesign: stopdesign.com/archive/2009/03/20/goodbye-google.html.

39. Alex Hern, "Why Google Has 200m Reasons to Put Engineers over Designers," *The Guardian*, February 5, 2014.

40. Marc Andreessen, Tweet, December 12, 2014 (since deleted). On September 24, 2016, Andreessen, whose handle on Twitter is @pmarca, announced that he was "taking a Twitter break!" and deleted more than 100,000 posts to Twitter, leaving that one tweet. The quotes from his Twitter account that appear here were copied by the author during research. Intriguingly, some 72,000 of his Tweets, starting in 2015, have been reposted by a Twitter account called "I retweet pmarca."

41. Sarah Lacy, *Once You're Lucky, Twice You're Good*, New York: Gotham Books, 2008, pp. 126–127.

42. Ben Horowitz, *The Hard Thing About Hard Things*, New York: Harper Collins, 2014.

43. Lessley Anderson, "Elon Musk: A Machine Tasked with Getting Rid of Spam Could End Humanity," *Vanity Fair*, October 8, 2014.

44. James Douglas, "Star Lords," *The Awl*, December 15, 2015.

45. To be precise, Google's tech workforce is 18 percent women and Facebook's is 17 percent women, while women hold 24 percent of Google's leadership positions and 27 percent of senior leadership positions at Facebook.

Amazon doesn't break down its tech workforce numbers, but the percentage of women in management positions was in the same area, 25 percent. The percentages of African Americans and Hispanics within Google's and Facebook's tech workforce were identical: 1 percent African American, 3 percent Hispanic. Uber recently released its first diversity numbers—tech workers were 15.4 percent women, 1 percent African American, and 2.1 percent Hispanic.

See Google Diversity Web site, accessed February 20, 2017: https://www.google.com/diversity/; Maxine Williams, "Facebook Diversity Update: Positive Hiring Trends Show Progress," July 14, 2016 Facebook Newsroom: https://newsroom.fb.com/news/2016/07/facebook-diversity-update-positive-hiring-trends-show-progress/; "Our Workforce Demographics," Diversity at Amazon, data as of July 2016: https://www.amazon.com/b?node=10080092011; Uber Diversity, data as of March 2017, https://www.uber.com/diversity/.

46. Robert Pogue Harrison, "The Children of Silicon Valley," *The New York Review of Books* blog, July 17, 2014.

47. Marc Andreessen, "Why Software Is Eating the World," *The Wall Street Journal*, August 20, 2011.

1. John McCarthy

1. Joseph Weizenbaum, *Computer Power and Human Reason: From Judgment to Calculation*, New York: W.H. Freeman and Company, 1976, pp. 226–227.

2. Weizenbaum, *Computer Power and Human Reason*, p. 227.

3. "Speaking Minds: Interviews with Twenty Eminent Cognitive Scientists," edited by Peter Baumgartner and Sabine Payr, Princeton, NJ: Princeton University Press, 1995, p. 253.

4. Ibid., p. 116.

5. John McCarthy, "An Unreasonable Book," appearing in Benjamin Kuipers, John McCarthy, and Joseph Weizenbaum, comments on *Computer Power and Human Reason*, in *ACM Sigart Newsletter* (Association for Computing Machinery, Special Interest Group on Artificial Intelligence) no. 58, June 1976, p. 8: http://www-formal.stanford.edu/jmc/reviews/weizenbaum.html.

6. Cheryl Zollars, "Scientists Discuss Role of Technology in Society," *The Stanford Daily*, May 19, 1978, p. 1.

7. John Markoff, interview with John McCarthy, July 19, 2002, personal copy.

8. There is no book-length biography of John McCarthy, and when asked later in life if he would write a memoir, he said he wouldn't because he was "not prepared to be honest" and address "what I could have accomplished if I

hadn't been lazy." However, he has given many retrospective interviews and written prolifically about his life. His extensive personal Web site, with links to papers, talks, essays, popular science articles, favorite sayings, and so on, is www-formal.stanford.edu/jmc/, with Stanford running its own site: http://jmc.stanford.edu/index.html.

A number of works describe McCarthy's life, including: Philip J. Hilts, *Scientific Temperaments: Three Lives in Contemporary* Science, New York: Simon and Schuster, 1982; John Markoff, *Machines of Loving Grace: The Quest for Common Ground Between Humans and Robots*, New York: HarperCollins, 2015, particularly chapter 4, "The Rise, Fall and Resurrection of A.I."; Nils J. Nilsson, *John McCarthy, 1927–2011: A Biographical Memoir*, Washington, D.C.: National Academy of Sciences, 2012, pp. 1–17; Patrick J. Hayes and Leora Morgenstern, "On John McCarthy's 80th Birthday, in Honor of His Contributions," *AI Magazine*, Winter 2007, pp. 93–102.

The oral histories are Nils Nilsson and John McCarthy, interview, September 12, 2007, Computer History Museum: http://www.computerhistory.org/collections/catalog/102658149; and William Aspray and John McCarthy, interview, March 2, 1989, Palo Alto, CA, Charles Babbage Institute, University of Minnesota, Minneapolis: purl.umn.edu/107476.

9. Markoff and McCarthy interview.

10. Kathryn Cullen-DuPont, *Encyclopedia of Women's History of America, Second Edition*, New York: Facts on File Inc., 2000, p. 184.

11. Susan McCarthy, "What Your Dentist Doesn't Want You to Know," part of Celebration of John McCarthy's Accomplishments, Stanford University, March 25, 2012: http://www.saildart.org/jmc2012.html.

12. M. Ilin, "100,000 Whys: A Trip Around the Room," translated by Beatrice Kinkead, Philadelphia: J.B. Lippincott Company, 1933, p. 9.

13. Markoff and McCarthy interview.

14. John McCarthy, "What Is Artificial Intelligence?," November 12, 2007, John McCarthy Web site: http://jmc.stanford.edu/artificial-intelligence/index.html.

15. Paul N. Edwards, *The Closed World: Computers and the Politics of Discourse in Cold War America* Cambridge, MA: MIT Press, 1996, p. 159.

16. "Hixon Symposium on Cerebral Mechanisms in Behavior," California Institute of Technology, Pacific State Hospital, Pomona, CA, September 20, 1948, accessed at Linus Pauling Day by Day Web Site, Oregon State University Libraries Special Collections: http://scarc.library.oregonstate.edu/coll/pauling/calendar/1948/09/20-xl.html.

17. Nilsson and McCarthy interview.

18. Markoff and McCarthy interview; Sylvia Nasar, *A Beautiful Mind: The Life of Mathematical Genius and Nobel Laureate John Nash*, New York: Simon and Schuster, 2001, p. 146.

19. Markoff and McCarthy interview.

20. See John McCarthy, "What Was Attractive About Marxism?," May 10, 2005: http://jmc.stanford.edu/commentary/progress/marxism2.ht ml; and John McCarthy, "Marxism," August 27, 2008: http://jmc.stanford .edu/commentary/progress/marxism.html.

21. Nilsson, *John McCarthy*, p. 4.

22. Nilsson and McCarthy interview.

23. J. McCarthy, M. L. Minsky, N. Rochester, C. E. Shannon, "A Proposal for the Dartmouth Summer Research Project on Artificial Intelligence," August 31, 1955, reprinted in *AI Magazine*, vol. 27, no. 4 (2006), pp. 12–14.

24. Aspray and McCarthy interview.

25. Margaret Hamilton, interview with author, October 2, 2015.

26. John McCarthy, "The Well-Designed Child," 1996: http://jmc .stanford.edu/articles/child.html.

27. John McCarthy, "Information," *Scientific American*, September 1966, pp. 65–72.

28. John McCarthy, "The Little Thoughts of Thinking Machines," *Psychology Today*, December 1983, pp. 46–49.

29. John McCarthy, "Artificial Intelligence and Creativity," Century 21 lecture, January 30, 1968, audio file in KZSU Collection, Stanford Archive of Recorded Sound, Stanford University Libraries.

30. John McCarthy, "An Example for Natural Language Understanding and the AI Problems It Raises," 1976: http://jmc.stanford.edu/articles/mrhug .html.

31. Nilsson, *John McCarthy*, p. 11.

32. Steven Levy, *Hackers: Heroes of the Computer Revolution*, New York: Penguin Books, 2001 (original 1984), p. 24.

33. Louis Fein, oral history interview conducted by Pamela McCorduck, May 9, 1979, Charles Babbage Institute, retrieved from the University of Minnesota Digital Conservancy: http://hdl.handle.net/11299/107284.

34. This paper was later published in an academic journal: Louis Fein, "The Role of the University in Computers, Data Processing, and Related Fields," *Communications of the ACM*, vol. 2, no. 9, September 1959, pp. 7–14.

35. Ibid., p. 13.

36. Quora, "How did Mark Zuckerberg become a programming prodigy," https://www.quora.com/How-did-Mark-Zuckerberg-become-a-programming-prodigy?redirected_qid=2886003.

37. Paul Graham, "Some Heroes," April 2006: http://paulgraham.com /heroes.html.

38. Levy, *Hackers*, pp. 26–27.

39. Tung-Hui Hu, *A Prehistory of the Cloud*, Cambridge, MA: MIT Press, 2015, p. 46.

40. Edwards, *The Closed World*, pp. 257–258.

41. Ibid., pp. 55–56.

42. Ibid.; see chapter 2, "The Hacker Ethic," pp. 39–49; quote appears on p. 67.

43. Ibid., p. 83.

44. Hamilton, author interview.

45. Levy, *Hackers*, pp. 40–49.

46. George E. Forsythe, "What to Do till the Computer Scientist Comes," Technical Report No. CS 77, Computer Science Department, School of Humanities and Sciences, Stanford University, September 22, 1967, p. 4.

47. Ibid.

48. McCarthy and Markoff interview, 2002.

49. Ibid.

50. Steven G. Ungar, "AI Goal: A 'Thinking' Machine," *The Stanford Daily*, January 26, 1971, p. 1.

51. John Markoff, "Optimism as Artificial Intelligence Pioneers Reunite," *The New York Times*, December 8, 2009, p. D4.

52. Bard Darrach, "Meet Shaky, the First Electronic Person," *Life*, November 20, 1970, p. 64.

53. Ungar, "AI Goal."

54. Classified advertisement, *The Stanford Daily*, May 25, 1971, p. 4.

55. Bruce Guenther Baumgart, "Saildart Prolegomenon 2016": www.saildart.org/book/0.pdf.

56. Stewart Brand, "Spacewar: Fanatic Life and Symbolic Death Among the Computer Bums," *Rolling Stone*, December 7, 1972.

57. Ungar, "AI Goal," p. 1.

58. John McCarthy, "The Home Information Terminal," appearing in "Man and Computer," Proceedings of the International Conference, Bordeaux, 1970, pp. 48–57 (Karger, Basel 1972).

59. John McCarthy, "The Home Information Terminal—a 1970 View," June 1, 2000: http://www-formal.stanford.edu/jmc/hoter2.pdf.

60. McCarthy and Markoff interview.

61. McCarthy, "The Home Information Terminal," p. 7.

62. McCarthy and Markoff interview.

63. Jim Wascher, "SRM Protest: Council Meeting Halted," *The Stanford Daily*, April 3, 1972.

64. Cheryl Zollars, "Scientists Discuss Role of Technology in Society," *The Stanford Daily*, May 19, 1978, p. 1.

65. John McCarthy, "Prophets—Especially Prophets of Doom," October 17, 1995: http://www-formal.stanford.edu/jmc/progress/prophets.html.

66. Lee Dembart, "Experts Argue Whether Computers Could Reason, and if They Should," *The New York Times*, May 8, 1977, p. A1.

67. John Markoff, "Joseph Weizenbaum, Famed Programmer, Is Dead at 85," *The New York Times*, March 13, 2008, p. A22.

68. Diana ben-Aaron, "Weizenbaum Examines Computers and Society," *The Tech*, vol. 105, no. 16, April 9, 1985, p. 2: http://tech.mit.edu/V105/N16/weisen.16n.html; Howard Rheingold, *Tools for Thought: The History and Future of Mind-Expanding Technology*, Cambridge, MA: The MIT Press, 2000, pp. 163–164.

69. Ben-Aaron, "Weizenbaum Examines Computers and Society."

70. Weizenbaum, *Computer Power and Human Reason*, p. 115.

71. The identity of the "patient" in this conversation is hard to pin down. In his academic paper on Eliza, Weizenbaum describes the conversation as "typical." See Joseph Weizenbaum, "ELIZA—A Computer Program for the Study of Natural Language Communication Between Man and Machine," *Communications of the ACM*, vol. 9, no. 1 (January 1966): 36–35. In *Computer Power and Human Reason*, however, he describes the computer's interlocutor as a "young woman." (Based on the dialogue, that appears to be the "part" being played.) In a *New York Times* account, which includes the same sample conversation, Weizenbaum is identified as typing in the statements himself: "In one test of Eliza's instructions, the following typewritten conversation took place between Mr. Weizenbaum, the patient (P.) and an IBM 7094 computer (C.) in the role of the doctor, with the latter 'unaware' of what the specific questions would be." John Noble Wilford, "Computer Is Being Taught to Understand English," *The New York Times*, June 15, 1968, p. 58.

72. Wilford, "Computer Is Being Taught to Understand English."

73. Joseph Weizenbaum, comments on *Computer Power and Human Reason*, in ACM Sigart Newsletter, p. 13.

74. "Speaking Minds: Interviews with Twenty Eminent Cognitive Scientists," edited by Peter Baumgartner and Sabine Payr, Princeton, NJ: Princeton University Press, 1995, p. 260.

75. Ibid., pp. 257–258.

76. *Plug and Pray*, directed by Jens Schanze, Mascha Films, 2010.

77. Joseph Weizenbaum and Reid Hoffman, "Virtual Worlds—Fiction or Reality?" Davos Open Forum 2008, January 26, 2008: https://www.youtube.com/watch?v=E198IynGbg0.

78. Jason Bloomstein, "Racial Slurs Cause University to Shut down Bulle-

tin Board," *The Stanford Daily*, January 30, 1989, p. 1. A summary of the controversy from McCarthy's perspective can be found on his Web site: "The Rec. Humor.Funny Censorship at Stanford University," May 12, 1996: http://jmc .stanford.edu/general/rhf.html. The digital archive at the Stanford Artificial Intelligence Lab preserved the emails within the Computer Science Department: http://www.saildart.org/FUNNY.89[BB,DOC].

79. See Daniel T. Rodgers, *The Age of Fracture*, Cambridge, MA: Harvard University Press, 2011, p. 210.

80. David Sacks and Peter Thiel, *The Diversity Myth: "Multiculturalism" and the Politics of Intolerance at Stanford*," Oakland, CA: The Independent Institute, 1995, p. xxi.

81. McCarthy, "The Rec.Humor.Funny Censorship at Stanford University."

82. John McCarthy email to su-etc@SAIL.Stanford.EDU and faculty@ SCORE.Stanford.EDU, January 29, 1989.

83. John McCarthy <JMC@SAIL.Stanford.EDU>, email to su-etc@ Sail.Stanford.edu, February 7, 1989: http://www.saildart.org/FUNNY .89[BB,DOC].

84. William Brown, Jr. <wab@sumex-aim.stanford.edu>, email to su-etc@sumex-aim.stanford.edu, February 14, 1989: http://www.saildart.org /FUNNY.89[BB,DOC].

85. Ibid.

86. Mark Crispin <mrc@sumex-aim.stanford.edu>, email to William Brown Jr, February 15, 1989: http://www.saildart.org/FUNNY.89[BB, DOC].

87. Andy Freeman <andy@gang-of-four.stanford.edu>, email to su-etc@ score.stanford.edu, February 15, 1989.

88. William Brown, Jr. <wab@sumex-aim.stanford.edu>, email to Andy Freeman, February 15, 1989.

89. W. Augustus Brown Jr., telephone interview with author, February 13, 2017.

90. John McCarthy <JMC@SAIL.Stanford.EDU>, email to J.JBREN NER@MACBETH.STANFORD.EDU, February 8, 1989: http://www .saildart.org/FUNNY.89[BB,DOC].

91. Oren Patashnik <op@polya.stanford.edu> email to su-etc@SAIL. Stanford.EDU and faculty@SCORE.Stanford.EDU et al., March 1, 1989.

92. John McCarthy, "Computer Science 40th Anniversary: A Symposium & Celebration Arrillaga Alumni Center," March 21, 2006: https://itunes .apple.com/us/itunes-u/department-computer-science/id385659431?mt=10.

93. McCarthy, "Computer Science 40th Anniversary."

2. Frederick Terman

1. Frederick Terman, letter to Paul Davis, December 1943, in Stuart W. Leslie, *The Cold War and American Science: The Military-Industrial-Academic Complex at MIT and Stanford*, New York: Columbia University Press, 1993, p. 44.

2. Rebecca S. Lowen, *Creating the Cold War University: The Transformation of Stanford*, Berkeley: University of California Press, 1997, p. 15.

3. Ibid., p. 148.

4. C. Stewart Gillmor, *Fred Terman at Stanford: Building a Discipline, a University, and Silicon Valley*, Stanford, CA: Stanford University Press, 2004, p. 491.

5. Carolyn E. Tajnai, "From the Valley of Heart's Delight to the Silicon Valley: A Study of Stanford University's Role in the Transformation," Stanford University, Department of Computer Science, 1996: http://forum .stanford.edu/carolyn/valley_of_hearts.

6. Ibid, p. 8.

7. Lowen, *Creating the Cold War University*, p. 173.

8. Gillmor, *Fred Terman at Stanford*, p. 379.

9. Lowen, *Creating the Cold War University*, p. 159.

10. Gillmor, *Fred Terman at Stanford*, p. 419.

11. Richard W. Lyman, *Stanford in Turmoil: Campus Unrest, 1966–1972*, Stanford, CA: Stanford University Press, 2009, p. 6.

12. Orrin Leslie Elliott, *Stanford University: The First Twenty Five Years*, Stanford, CA: Stanford University Press, 1937, pp. 16–17: https://archive.org /details/stanfroduniversi009361mbp.

13. Ibid., p. 12.

14. Elliott, *Stanford University*, pp. 15–16.

15. Stanford Facts 2016, "The Campus Plan," http://facts.stanford.edu /about/lands.

16. Gillmor, *Fred Terman at Stanford*, p. 17.

17. Stanford University, "The Founding Grant with Amendments, Legislation, and Court Decrees," 1987, p. 24.

18. Elliott, *Stanford University*, p. 39.

19. Ibid., pp. 21–22.

20. Richard Hofstadter and Walter P. Metzger, *The Development of Academic Freedom in the United States*, New York: Columbia University Press, 1955, p. 413.

21. Ibid., p. 414.

22. Elliott, *Stanford University*, pp. 50–51.

23. Stanford University, "Birth of a University": https://www.stanford.edu/about/history/.

24. W. B. Carnochan, "The Case of Julius Goebel: Stanford, 1905," *The American Scholar*, vol. 72, no. 3 (Summer 2003): p. 95.

25. Margo Davis and Roxanne Nilan, *The Stanford Album: A Photographic History, 1885–1945*, Stanford, CA: Stanford University Press, 1989, p. 47.

26. Elliott, *Stanford University*, p. 270.

27. Ibid., p. 272.

28. Hofstadter and Metzger, *The Development of Academic Freedom in the United States*, p. 436.

29. Elliott, *Stanford University*, "The 500 Limit," pp. 132–136.

30. Stanford University, Founding Grant. The cap at five hundred women was first reached in 1903 and continued until 1933, when it was revised to a 40 percent quota, which reflected the ratio in 1899, when Mrs. Stanford made her change to the founding grant. In 1973, Stanford petitioned a court to have the grant amended to remove any limits on the number of women attending the university.

31. Elliott, *Stanford University*, p. 132.

32. Ibid., p. 135.

33. Elliott, *Stanford University*, p. 160.

34. Susan Wolfe, "Who Killed Jane Stanford?" *Stanford Alumni* magazine, September/October 2003: https://alumni.stanford.edu/get/page/magazine/article/?article_id=36459.

35. Carnochan, "The Case of Julius Goebel," p. 108.

36. Wolfe, "Who Killed Jane Stanford?"

37. Peter Hegarty, *Gentlemen's Disagreement: Alfred Kinsey, Lewis Terman, and the Sexual Politics of Smart Men*, Chicago: University of Chicago Press, 2013, p. 3.

38. See John D. Wasserman, "A History of Intelligence Assessment: The Unfinished Tapestry," in *Contemporary Intellectual Assessment, Third Edition*, edited by Dawn P. Flanagan and Patti L. Harrison, New York: The Guilford Press, 2012.

39. Ibid., "Lewis Terman: Trails to Psychology," pp. 297–331.

40. B. R. Hergenhahn and Tracy Henley, *An Introduction to the History of Psychology, Seventh Edition*, Belmont, CA: Cengage Learning, 2013, p. 304. The full title of Terman's dissertation: "Genius and Stupidity: A Study of the Intellectual Processes of Seven 'Bright' and Seven 'Stupid' Boys."

41. Wasserman, "A History of Intelligence Assessment," p. 17.

42. Henry L. Minton, *Lewis M. Terman: Pioneer in Psychological Testing*, New York: New York University Press, 1988.

43. Ibid., p. 20.

44. Alfred Binet and Théodore Simon, *The Development of Intelligence in Children*, Baltimore: Wilkins & Wilkins Company, 1916, pp. 42–43.

45. L. M. Terman, "Were We Born That Way?" *World's Work*, October 1922, vol. 44, no. 6, p. 659.

46. Edwin G. Boring, "Lewis Madison Terman (1877–1956)," Washington, D.C.: National Academy of Sciences, 1959, p. 429.

47. Daniel Goleman, "75 Years Later, Study Still Tracking Geniuses," *The New York Times*, March 7, 1995, p. C1.

48. Mitchell Leslie, "The Vexing Legacy of Lewis Terman," *Stanford Magazine*, July/August 2000: https://alumni.stanford.edu/get/page/magazine/article/?article_id=40678.

49. See Adam Cohen, *Imbeciles: The Supreme Court, American Eugenics, and the Sterilization of Carrie Buck*, New York: Penguin, 2016.

50. Walter Lippmann, "The Abuse of the Tests," *The New Republic*, November 15, 1922, p. 297.

51. Lewis M. Terman, "The Great Conspiracy, or the Impulse Imperious of Intelligence Testers, Psychoanalyzed and Exposed by Mr. Lippmann," *The New Republic*, December 27, 1922, p. 119.

52. Mitchell Leslie, "The Vexing Legacy of Lewis Terman."

53. Boring, "Lewis Madison Terman," p. 440.

54. See Gillmor, *Fred Terman at Stanford*, chapter 1, "California Boy," pp. 11–69.

55. Stuart W. Leslie, *The Cold War and American Science*, p. 49.

56. Gillmor, *Fred Terman at Stanford*, p. 90.

57. Stuart W. Leslie, *The Cold War and American Science*, p. 53

58. Gillmor, *Fred Terman at Stanford*, p. 251

59. Lowen, *Creating the Cold War University*, pp. 130–131.

60. Lyman, *Stanford in Turmoil*, p. 9.

61. Lowen, *Creating the Cold War University*, p. 135.

62. Quoted in Lowen, *Creating the Cold War University*, p. 137.

63. Gillmor, *Fred Terman at Stanford*, p. 341.

64. Gillmor, *Fred Terman at Stanford*, p. 333.

65. Lowen, *Creating the Cold War University*, pp. 14, 104.

66. Lyman, *Stanford in Turmoil*, p. 8.

67. Nils Nilsson and John McCarthy interview, September 12, 2007, Computer History Museum: http://www.computerhistory.org/collections/catalog/102658149.

68. Allison Tracy and Edward Albert Feigenbaum interview, Stanford Historical Society Oral History Program, 2012, p. 20.

69. Stanford News, "Neuroscience Pioneer Marc Tessier-Lavigne Named Stanford's Next President," February 4, 2016: http://news.stanford.edu /features/2016/president-named/.

70. Stanford University, "OTL and the Inventor: Roles in Technology Transfer," retrieved October 22, 2016: http://otl.stanford.edu/inventors /resources/inventors_otlandinvent.html.

71. Lisa M. Krieger, "Stanford Earns $336 Million off Google Stock," *The San Jose Mercury News*, December 1, 2005, p. A1.

72. Stanford News, "Jen-Hsun Huang Pledges $30 Million for Innovative Engineering Center at Stanford," September 10, 2008: http://news.stanford .edu/pr/2008/pr-building-092408.html.

73. See Stanford University, "Fact Sheet: Sustainable Demolition— Frederick E. Terman Engineering Center": http://sustainable.stanford.edu /sites/default/files/documents/FactSheet_SustainableDemolition.pdf.

74. Kathleen J. Sullivan, "Excavator Tears down Walls, Ceilings and Floors of Terman Engineering Center," Stanford News, October 18, 2011.

3. Bill Gates

1. David K. Allison, "Transcript of a Video History Interview with Mr. William 'Bill' Gates," National Museum of American History, Smithsonian Institution: http://americanhistory.si.edu/comphist/gates.htm. "A major milestone for us was when we were walking through Harvard Square, one time, and saw this *Popular Electronics* magazine. And it was kind of in a way, good news and bad news. Here was someone making a computer around this chip in exactly the way that Paul had talked to me, and we'd thought about what kind of software could be done for it, and it was happening without us."

2. Walter Isaacson, "Dawn of a Revolution," *The Harvard Gazette*, September 20, 2013.

3. Stephen Manes and Paul Andrews, *Gates: How Microsoft's Mogul Reinvented an Industry—and Made Himself the Richest Man in America*, New York: Touchstone, 1994.

4. James Wallace and Jim Erickson, *Hard Drive: Bill Gates and the Making of the Microsoft Empire*, New York: Harper Collins, 1993, pp. 46–47.

5. Allison and Gates interview.

6. Ibid.

7. Ibid., p. 33.

8. Isaacson, "Dawn of a Revolution."

9. See Steven Levy, *Hackers: Heroes of the Computer Revolution*, Penguin Books, New York, 2001 (original 1984), pp. 224–243.

10. Allison and Gates interview.

11. Manes and Andrews, *Gates*, p. 26.

12. Daniel Golden and John Yemma, "Harvard Amasses a Colossal Endowment," *The Boston Globe*, May 31, 1988, p. A1.

13. Allison and Gates interview.

14. Scott Malone, "Dropout Bill Gates Returns to Harvard for Degree," Reuters, June 7, 2007: http://www.reuters.com/article/us-microsoft-gates -idUSN0730259120070607.

15. William Henry Gates III, "An Open Letter to Hobbyists," Homebrew Computer Club Newsletter, vol. 2, no. 1, January 31, 1976, p. 2.

16. *Homebrew Computer Club Newsletter*, vol. 1, no. 1, March 15, 1975.

17. Y Combinator, "Mark Zuckerberg at Startup School 2013" (2013), Zuckerberg Transcripts, Paper 160: http://dc.uwm.edu/zuckerberg_files _transcripts/160.

18. Bill Gates, @billgates, Twitter, May 15, 2017.

4. Marc Andreessen

1. Computer science's roots in artificial intelligence were a persistent concern for IBM, according to John McCarthy. "IBM thought that artificial intelligence was bad for IBM's image—that machines that were as smart as people and so forth were bad for their image. This may have been associated with one of their other image slogans, which was 'data processing, not computing.' That is, they were trying to get computing into business, so they wanted it to look as familiar and unfrightening as possible." John McCarthy, oral history interview with William Aspray, 1989, Charles Babbage Institute.

2. See Matthew Lyon and Katie Hafner, *Where Wizards Stay up Late: The Origins of the Internet*, New York: Simon & Schuster, 1999.

3. Tim Berners-Lee with Mark Fischetti, *Weaving the Web: The Original Design and Ultimate Destiny of the World Wide Web*, New York: Harper, 2000, pp. 12–13.

4. Tim Berners-Lee, "Frequently Asked Questions," on personal home page hosted by the World Wide Web Consortium (W3C): https://www.w3 .org/People/Berners-Lee/FAQ.html.

5. See Tad Friend, "Tomorrow's Advance Man," *The New Yorker*, May 18, 2015.

6. Ben Horowitz, annotations to Tad Friend, "Tomorrow's Advance Man," at Genius.com: http://genius.com/summary/www.newyorker.com %2Fmagazine%2F2015%2F05%2F18%2Ftomorrows-advance -man?unwrappable=1.

7. Joshua Quittner and Michelle Slatalla, *Speeding the Net: The Inside Story*

of Netscape and How It Changed Microsoft, New York: Atlantic Monthly Press, 1998, pp. 12–13.

8. Marc Andreessen, Tweet, May 10, 2015.

9. Quittner and Slatalla, *Speeding the Net*, p. 14.

10. James Romenesko, "Netscape's Wonder Boy Says of His Wisconsin Life: Oh, Yuck!" St. Paul Pioneer Press, June 1, 1998, p. 3E.

11. Matt Beer, "Net Scope," *The San Francisco Examiner*, May 23, 1999.

12. Friend, "Tomorrow's Advance Man."

13. Marc Andreessen, Tweet, September 4, 2015.

14. Quittner and Slatalla, *Speeding the Net*, p. 15.

15. Beer, "Net Scope."

16. Ibid.

17. The National Science Foundation, "Cyberinfrastructure: From Super-computing to the TeraGrid": https://www.nsf.gov/news/special_reports/cyber/fromsctotg.jsp.

18. Quittner and Slatalla, *Speeding the Net*, pp. 9–15.

19. Julie Bort, "Marc Andreessen Gets All the Credit for Inventing the Browser but This Is the Guy Who Did 'All the Hard Programming,'" *Business Insider*, May 12, 2014.

20. David A. Kaplan, "Nothing but Net," *Newsweek*, December 25, 1995, p. 32.

21. Marc Andreessen, "New 'XMosaic' World-Wide Web Browser from NCSA," January 23, 1993, message forwarded by Tim Berners-Lee to news-groups alt.hypertext and comp.infosystems: http://www.bio.net/bionet/mm/bio-soft/1993-February/003879.html.

22. David K. Allison, "Excerpts from an Oral History Interview with Marc Andreessen," Smithsonian Institution, June 1995: http://americanhistory.si.edu/comphist/ma1.html.

23. John Markoff, "A Free and Simple Computer Link," December 8, 1993, p. D1.

24. Berners-Lee, *Weaving the Web*, p. 68.

25. Scott Laningham, "DeveloperWorks Interviews: Tim Berners-Lee," IBM, August 22, 2006: http://www.ibm.com/developerworks/podcast/dwi/cm-int082206txt.html.

26. Walter Isaacson, *The Innovators: How a Group of Hackers, Geniuses, and Geeks Created the Digital Revolution*, New York: Simon and Schuster, 2014, p. 417.

27. Berners-Lee, *Weaving the Web*, pp. 70–71.

28. John C. Thomson Jr., "Privatization of the New Communication Channel: Computer Networks and the Internet," Web paper on Internet privatization, Fall 2000: http://johnthomson.org/j561/index.html.

29. Ibid.

30. See Paul Andrews, "Profit Without Honor," *The Seattle Times*, October 5, 1997.

31. Laningham, "DeveloperWorks Interviews: Tim Berners-Lee."

32. Berners-Lee, *Weaving the* Web, p. 57.

33. Isaacson, *The Innovators*, p. 417.

34. Ibid.

35. Brad Stone, *The Everything Store: Jeff Bezos and the Age of Amazon*, New York: Little, Brown, 2013, p. 25.

36. Quittner and Slatalla, *Speeding the Net*, p. 74.

37. Allison, "Oral History Interview with Marc Andreessen."

38. John Markoff, "A Free and Simple Computer Link," *The New York Times*, December 8, 1993, p. D1.

39. Michael Lewis, *The New New Thing: A Silicon Valley Story*, New York: W. W. Norton, 2000, p. 81.

40. Andrews, "Profit Without Honor."

41. Kaplan, "Nothing but Net."

42. Adam Lashinsky, "Remembering Netscape: The Birth of the Web," *Fortune*, July 25, 2005.

43. Lewis, *The New New Thing*, p. 40.

44. Andrews, "Profit Without Honor."

45. Lou Montulli, interviewed by Brian McCullough, Internet History Podcast, episode 5, March 6, 2014: http://www.internethistorypodcast.com /2014/03/chapter-1-supplemental-1-an-interview-with-lou-montulli/.

46. See Jamie Zawinski, "The Condensed and Expurgated History of the About:authors URL": https://www.jwz.org/doc/about-authors.html. An early feature of the Netscape browser was the address about:authors, which would take users to a page listing the programming team responsible for the browser. The passage continues, "Thus it amounts to the same thing whether one gets drunk alone or is a leader of nations. If one of these activities takes precedence over the other . . . it will be the quietism of the solitary drunkard which will take precedence over the vain agitation of the leader of nations." Jean-Paul Sartre, *Being and Nothingness*, translated by Hazel E. Barnes, New York: Washington Square Press, 1993, p. 797.

47. Matthew Gray, "Measuring the Growth of the Web: June 1993 to June 1995," 1995: http://www.mit.edu/people/mkgray/growth/.

48. W. Joseph Campbell, "The '90s Startup That Terrified Microsoft and

Got Americans to Go Online," *Wired*, January 27, 2015: https://www.wired.com/2015/01/90s-startup-terrified-microsoft-got-americans-go-online/.

49. Marc Andreessen, Internet Gazette Multimedia Conference, San Francisco, November 5, 1994.

50. Ibid.

51. Allison, "Oral History Interview with Marc Andreessen."

52. Andrews, "Profit Without Honor."

53. Marc Andreessen, Twitter, January 12, 2014.

54. Andrews, "Profit Without Honor."

55. Andreessen, Internet Gazette Multimedia Conference.

56. Berners-Lee, *Weaving the Web*.

57. Lou Montulli, "The Reasoning Behind Web Cookies," The Irregular Musings of Lou Montulli Blog, May 14, 2013: http://www.montulli-blog.com/2013/05/the-reasoning-behind-web-cookies.html.

58. See Rajiv C. Shah and Jay P. Kesan, "Recipes for Cookies: How Institutions Shape Communication Technologies" (July 15, 2004), *New Media & Society*, May 1, 2009, vol. 11, no. 3: 315–336.

59. Lou Montulli, interviewed by Brian McCullough.

60. See Shah and Kesan, "Recipes for Cookies."

61. See user profile of blue_beetle, aka Andrew Lewis, who has made T-shirts featuring the phrase: http://www.metafilter.com/user/15556.

62. Tim Jackson, "This Bug in Your PC Is a Smart Cookie," *Financial Times*, February 12, 1996, p. 15.

63. Lee Gomes, "Web 'Cookies' May Be Spying on You," *The San Jose Mercury News*, February 13, 1996.

64. Montulli, "The Reasoning Behind Web Cookies." See also Lou Montulli, "Why Blocking 3rd Party Cookies Could Be a Bad Thing," The Irregular Musings of Lou Montulli Blog, May 17, 2013: http://www.montulli-blog.com/2013/05/why-blocking-3rd-party-cookies-could-be.html.

65. Lashinsky, "Remembering Netscape."

66. Ibid.

67. Ted Greenwald, "How Jimmy Wales' Wikipedia Harnessed the Web as a Force for Good," *Wired*, March 19, 2013.

68. Lashinsky, "Remembering Netscape."

69. Estimates vary about market share, but one reliable tracker at the time based at the University of Illinois, Urbana-Champaign, came up with these figures. Accessed via the Wayback Machine, which takes snapshots of Web sites: https://web.archive.org/web/20010507150536, http://www.ews.uiuc.edu/bstats/months/9710-month.html.

70. Campbell, "The '90s Startup."

71. Quittner and Slatalla, *Speeding the Net*, pp. 76–77.

72. John Perry Barlow, "A Declaration of the Independence of Cyberspace," February 8, 1996: https://www.eff.org/cyberspace-independence.

73. Lashinsky, "Remembering Netscape."

74. Kaplan, "Nothing but Net," p. 32.

75. Marc Andreessen, "Why Software Is Eating the World," *The Wall Street Journal*, August 20, 2011.

76. Todd Rulon-Miller, who became Netscape's vice president of sales, described his interview with Andreessen, who was twenty-three at the time. "He was on a workstation staring intently into the screen. I don't think he looked at me. I sat in a chair next to him. He was playing Doom." In Lashinksy, "Remembering Netscape."

77. Friend, "Tomorrow's Advance Man."

78. Mary Anne Ostrom, "From Peak to Valley; Andreessen Offers Perspective on Good Times and Hard Times in Tech," *The San Jose Mercury News*, March 6, 2003, 1C.

79. Marc Andreessen, posts to Twitter, July 13, 2014.

80. Sam Biddle, "Deep Thoughts with Marc Andreessen: The Poor Have It Pretty Good!," *ValleyWag*, June 4, 2014.

81. Maya Kosoff, "Marc Andreessen Quit Twitter and Now He Feels 'Free as a Bird,'" VanityFair.com, September 30, 2016.

82. Berners-Lee, *Weaving the Web*, p. 2.

83. Ibid., pp. 107–108.

84. Ibid., pp. 30–31.

5. Jeff Bezos

1. Marc Andreessen, Internet Gazette Multimedia Conference, San Francisco, November 5, 1994.

2. Charles King, "1995: The Year the Internet Transformed Business," Pund-It blog, July 15, 2015: http://www.pund-it.com/blog/1995-the-year -the-internet-transformed-business/.

3. Jessica Livingston, *Founders at Work: Stories of Startups' Early Days*, Berkeley, CA: Apress, 2008, pp. 248–249.

4. Craigslist, "About: Expansion": https://www.craigslist.org/about /expansion.

5. Livingston, *Founders at Work*, p. 250.

6. Jessica Mintz, "Craigslist Chief Executive Tells Investment Community Users, not Money, Drives Business," Associated Press, December 7, 2006.

7. See Adam Cohen, *The Perfect Store: Inside eBay*, New York: Little, Brown, 2003.

8. Ibid., p. 25.

9. Ibid., p. 76.

10. Our History, eBay Web site: https://www.ebayinc.com/our-company /our-history/.

11. Cohen, *The Perfect Store*, p. 76.

12. Randall Stross, *eBoys: The First Inside Account of Venture Capitalists at Work*, New York: Crown Publishing Group, 2001, p. xv.

13. Peter de Jonge, "Riding the Wild, Perilous Waters of Amazon.com," *The New York Times Magazine*, March 14, 1999.

14. Les Earnest, "Automatic Investments," SAIL Sagas, December 13, 2009: https://web.stanford.edu/~learnest/spin/sagas.htm.

15. David Shaw, email re: 2009 SAIL Reunion, October 29, 2009, accessed via Regrets Web page: https://web.stanford.edu/~learnest/spin/regrets.htm.

16. "Biophysicist in Profile: David E. Shaw," Biophysical Society Newsletter, January 2016, pp. 2–3: https://biophysics.cld.bz/Biophysical-Society -Newsletter-January-2016/2#2/z.

17. James Aley, "Wall Street's King Quant David Shaw's Secret Formulas Pile up Money. Now He Wants a Piece of the Net," *Fortune*, February 5, 1996.

18. Michael Peltz, "The Power of Six," Institutional Investors' Alpha, March 2009.

19. Stone, *The Everything Store*, p. 27.

20. Aley, "Wall Street's King Quant."

21. Ibid.

22. Joel Shurkin, *Broken Genius: The Rise and Fall of William Shockley, Creator of the Electronic Age*, New York: Macmillan, 2006, p. 164.

23. C. Stewart Gillmor, *Fred Terman at Stanford: Building a Discipline, a University, and Silicon Valley*, Stanford, CA: Stanford University Press, 2004, p. 11.

24. See "The High Cost of Thinking the Unthinkable," in Shurkin, *Broken Genius*, pp. 241–256.

25. Stone, *The Everything Store*, p. 20.

26. Jeff Bezos, "We Are What We Choose," Baccalaureate Remarks at Princeton University, May 30, 2010: http://www.princeton.edu/main/news/ archive/S27/52/51O99/index.xml.

27. De Jonge, "Riding the Wild, Perilous Waters of Amazon.com."

28. Department of Computer Science Alumni News, University of Illinois, Urbana-Champaign, Summer 2001, vol. 2, no. 6, p. 10: http://www.cs .uiuc.edu/sites/default/files/newsletters/summer01.pdf.

29. Craig Cannon, "Employee #1: Amazon," Y Combinator blog, September 6, 2016: http://blog.ycombinator.com/employee-1-amazon/.

30. Ibid.

31. John Cook, "Meet Amazon.com's First Employee: Shel Kaphan," *GeekWire*, June 14, 2011: http://www.geekwire.com/2011/meet-shel-kaphan -amazoncom-employee-1/.

32. Stone, *The Everything Store*, p. 40.

33. Stone, *The Everything Store*, p. 37.

34. Cook, "Meet Amazon.com's First Employee."

35. Cannon, "Employee #1."

36. De Jonge, "Riding the Wild, Perilous Waters of Amazon.com."

37. Tonya Garcia, "Amazon Will Account for More Than Half of 2015 E-commerce Growth, Says Macquarie," *MarketWatch*, December 22, 2015: http://www.marketwatch.com/story/amazon-will-account-for-more-than -half-of-2015-e-commerce-growth-says-macquarie-2015-12-22.

38. "Amazon.com Announces Fourth Quarter Sales up 22% to $35.7 Billion," *BusinessWire*, January 28, 2016: http://www.businesswire.com/news /home/20160128006357/en/Amazon.com-Announces-Fourth-Quarter -Sales-22-35.7.

39. Jodi Kantor and David Streitfeld, "Amazon's Bruising, Thrilling Workplace," *The New York Times*, August 16, 2015, p. A1.

40. See, e.g., Spencer Soper, "Inside Amazon's Warehouse," *The Morning Call* (Allentown, PA), September 18, 2011; Spencer Soper, "Amazon Workers Left out in the Cold," *The Morning Call* (Allentown, PA), November 6, 2011; Hamilton Nolan, "True Stories of Life as an Amazon Worker," *Gawker*, August 2, 2013: http://gawker.com/true-stories-of-life-as-an-amazon -worker-1002568208.

41. Jeff Bezos, "Amazon Chief's Message to Employees," *The New York Times*, August 18, 2015, p. B3.

6. Sergey Brin and Larry Page

1. Sergey Brin and Lawrence Page, "The Anatomy of a Large-Scale Hypertextual Web Search Engine," Seventh International World-Wide Web Conference (WWW 1998), April 14–18, 1998, Brisbane, Australia.

2. Marc Andreessen, Internet Gazette Multimedia Conference, San Francisco, November 5, 1994.

3. Ibid.

4. Katherine Losse, *The Boy Kings: A Journey into the Heart of the Social Network*, New York: Free Press, 2012, p. 6.

5. Steven Levy, *In the Plex: How Google Thinks, Works, and Shapes Our Lives*, New York: Simon and Schuster, 2011, pp. 19–20.

6. Brin and Page, "The Anatomy of a Large-Scale Hypertextual Web Search Engine."

7. Christopher Lyle and Ravi Sarin, "Googles of Dollars," *The Stanford Daily*, April 11, 2000, p. B1.

8. Levy, *In the Plex*, p. 21.

9. Brin and Page, "The Anatomy of a Large-Scale Hypertextual Web Search Engine."

10. Lawrence Page, Sergey Brin, Rajeev Motwani, and Terry Winograd, "The PageRank Citation Ranking: Bringing Order to the Web," Technical Report (1999), Stanford InfoLab, pp. 11–12: http://ilpubs.stanford.edu:8090 /422/.

11. Levy, *In the Plex*, p. 35.

12. John Markoff, interview with John McCarthy, July 19, 2002, personal copy.

13. Brin and Page, "The Anatomy of a Large-Scale Hypertextual Web Search Engine," p. 3.

14. Ibid.

15. Ibid., p. 18.

16. Ibid., p. 19.

17. Google Inc. press release: "Google Receives $25 Million in Equity Funding: Sequoia Capital and Kleiner Perkins Lead Investment; General Partners Michael Moritz and John Doerr Join Board," June 7, 1999: http:// googlepress.blogspot.com/1999/06/google-receives-25-million-in-equity.html.

18. Stanford School of Engineering Web site, "Gates Computer Science Building": https://www-cs.stanford.edu/about/gates-computer-science -building.

19. Levy, *In the Plex*, p. 23.

20. Eli Pariser, *The Filter Bubble: How the New Personalized Web Is Changing What We Read and How We Think*, New York: Penguin, 2011, p. 31.

21. Levy, *In the Plex*, p. 27.

22. Ibid., p. 13.

23. David K. Allison, "Excerpts from an Oral History Interview with Marc Andreessen," Smithsonian Institution, June 1995: http://americanhistory.si .edu/comphist/ma1.html.

24. John F. Ince, "The Lost Google Tapes: Conversations Tape-Recorded in the Early Years with Google's Founders Illuminate How Their Actions Forged the Growth of a Silicon Valley Giant," *The San Francisco Chronicle*, December 3, 2006.

25. Jessica Livingston, *Founders at Work: Stories of Startups' Early Days*, Berkeley, CA: Apress, 2008, p. 128.

26. Ibid., p. 134.

27. Ibid.

28. Peter Sinton, "Sequoia Grows Ventures," *The San Francisco Chronicle*, April 17, 1996, p. B1.

29. David F. Salisbury, "Yahoo! Founders Endow New Stanford Chair," Stanford News, February 12, 1997: http://news.stanford.edu/pr/97/970212 yahoo.html.

30. Mark Shwartz, "Alumni Couple Yang and Yamazaki Pledge $75 Million to the University," Stanford News, February 15, 2007: http://news .stanford.edu/news/2007/february21/donors-022107.html.

31. Ince, "The Lost Google Tapes."

32. Levy, *In the Plex*, p. 33.

33. Jacob Jolis, "Frugal After Google," *The Stanford Daily*, April 16, 2010, p. 3.

34. Levy, *In the Plex*, p. 34; and Jolis, "Frugal After Google."

35. Ince, "The Lost Google Tapes."

36. Ken Auletta, "Googled: The End of the World as We Know It," New York: Penguin Press, 2009, p. 44.

37. Levy, *In the Plex*, p. 74.

38. Ince, "The Lost Google Tapes."

39. See Levy, *In the Plex*, pp. 136–142.

40. Auletta, "Googled," p. 56.

41. Levy, *In the Plex*, p. 79.

42. Auletta, "Googled," pp. 66–68.

43. Levy, *In the Plex, Part Two: Googlenomics*, pp. 69–120.

44. Stefanie Olsen, "Google Files for Unusual $2.7 billion IPO," CNet, April 30, 2004.

45. Brin and Page, "The Anatomy of a Large-Scale Hypertextual Web Search Engine."

46. Harvard University, "CS50 Guest Lecture by Mark Zuckerberg" (2005), Zuckerberg Transcripts, Paper 141: http://dc.uwm.edu /zuckerberg_files_transcripts/141.

47. Lisa M. Krieger, "Stanford Earns $336 Million off Google Stock," *The San Jose Mercury News*, December 1, 2005, p. A1.

48. The Thiel Fellowship Web site home page: http://thielfellowship.org/.

7. Peter Thiel

1. For Thiel biography, see George Packer, *The Unwinding: An Inner History of the New America*, New York: Farrar, Straus and Giroux, 2013, pp. 120–136, 209–216, 381–397.

2. Stanford Technology Ventures Program, Thiel and Levchin, "Entrepreneurial Thought Leader Speaker Series," January 21, 2004.

3. Max Levchin and Peter Thiel, "PayPal Cofounders Met in Terman at a Seminar," Stanford eCorner, January 21, 2004: http://ecorner.stanford.edu /videos/1022/Paypal-Cofounders-Met-in-Terman-at-a-Seminar.

4. Eric M. Jackson, *The PayPal Wars: Battles with eBay, the Media, the Mafia, and the Rest of Planet Earth*, Washington, D.C.: WND Books, 2012, p. 5.

5. "A Conversation with Max Levchin," *The Charlie Rose Show*, August 1, 2013.

6. Jessica Livingston, *Founders at Work: Stories of Startups' Early Days*, New York: Apress, 2008, pp. 1–16.

7. Jackson, *The PayPal Wars*, p. 19.

8. Peter Thiel, "Address to Republican National Convention," Cleveland, OH, July 21, 2016, C-Span: https://www.c-span.org/video/?c4612796 /peter-thiel-addresses-republican-national-convention-proud-gay.

9. Packer, *The Unwinding*, pp. 120–121

10. Peter Thiel with Blake Masters, *Zero to One: Notes on Startups, or How to Build the Future*, New York: Penguin Random House, 2014, p. 37.

11. Isaac Barchas, "The Voice of the Right," *The Stanford Daily*, February 23, 1989, p. 4.

12. David O. Sacks and Peter A. Thiel, *The Diversity Myth: "Multiculturalism" and the Politics of Intolerance at Stanford*, Oakland, CA: The Independent Institute, 1995, p. xxi.

13. Ibid., p. 152

14. Ibid., p. 140.

15. Ibid., p. 238.

16. Brad Hayward, "Report Explores Motives of Frosh," *The Stanford Daily*, January 18, 1989, p. 1.

17. Sacks and Thiel, *The Diversity Myth*, p. 41.

18. Ibid.

19. Brad Hayward, "Ujamaa Case Ends Without Charges," *The Stanford Daily*, February 10, 1989, p. 1.

20. See Sacks and Thiel, *The Diversity Myth*, pp. 169–174.

21. Keith Rabois, "Rabois: My Intention Was to Make a Provocative Statement," *The Stanford Daily*, February 7, 1992, p. 5.

22. Juthymas Harntha, "Two Students Can't Be Charged for Hurling Homophobic Slurs," *The Stanford Daily*, February 3, 1992, p. 1.

23. See Sacks and Thiel, *The Diversity Myth*, p. 174.

24. Jackson, *The PayPal Wars*, p. 204.

25. Sacks and Thiel, *The Diversity Myth*, p. 147.

26. Owen Thomas, "Peter Thiel Is Totally Gay, People," *Gawker*, December 19, 2007: http://gawker.com/335894/peter-thiel-is-totally-gay-people.

27. Connie Loizos, "Peter Thiel on Valleywag; It's the 'Silicon Valley Equivalent of Al Qaeda,'" *PE Hub*, May 18, 2009.

28. Jeffrey Toobin, "Gawker's Demise and the Trump-Era Threat to the First Amendment," *The New Yorker*, December 19–26, 2016.

29. Ibid. and Andrew Ross Sorkin, "Tech Billionaire in a Secret War with Gawker," *The New York Times*, p. A1, May 26, 2016.

30. Julia Carrie Wong, "Peter Thiel, Who Gave $1.25m to Trump, Has Called Date Rape 'Belated Regret,'" *The Guardian*, October 21, 2016.

31. Ryan Mac and Matt Drange, "Donald Trump Supporter Peter Thiel Apologizes for Past Book Comments on Rape," Forbes.com, October 25, 2016.

32. Sacks and Thiel, *The Diversity Myth*, p. 59.

33. Ibid., pp. 58–59.

34. "Undergraduate Senators," *The Stanford Daily*, April 9, 1987, p. 6.

35. Katie Mauro, "4 Groups Denied Spot on Fee-Request Ballot," *The Stanford Daily*, February 27, 1992, p. 1.

36. Elise Wolfgram, "Groups Cry Foul on Granting of Funds to Review," *The Stanford Daily*, May 29, 1992, p. 1.

37. Thiel, *Zero to One*, pp. 36–37.

38. Peter Thiel on the Future of Innovation with Tyler Cowen, "Conversations with Tyler," Mercatus Center, April 6, 2015: https://medium.com/conversations-with-tyler/peter-thiel-on-the-future-of-innovation-77628a43c0dd#.bav03wzih.

39. Kevin Wacknov, "Championship Stanford Chess Team Never Board," *The Stanford Daily*, December 2, 1986, p. 7.

40. Thiel, *Zero to One*, p. 141.

41. "A Conversation with Max Levchin," *The Charlie Rose Show*.

42. Peter Thiel, *Zero to One*, p. 173.

43. See PayPal slide show at Max Levchin Web site: July 1999 staff photo: http://levchin.com/paypal-slideshow/2.html.

44. Thiel, *Zero to One*, p. 122.

45. Jackson, *The PayPal Wars*, p. 19.

46. Ibid., p. 92.

47. Peter Thiel with Tyler Cowen, "Conversations With Tyler."

48. Levchin and Thiel, Stanford eCorner.

49. Jackson, *The PayPal Wars*, p. 33.

50. Ibid., pp. 44–46.

51. Levchin and Thiel, Stanford eCorner.

52. See Jackson, *The PayPal Wars*, passim 51–126.

53. Livingston, *Founders at Work*, p. 10.

54. Levchin and Thiel, Stanford eCorner.

55. Jackson, *The PayPal Wars*, p. 152.

56. Max Levchin, "Data and Risk," DLD13 Keynote, January 21, 2013.

57. Bloomberg News, "EBay Settles Suit by PayPal Customers," *The Los Angeles Times*, June 15, 2004, p. C5.

58. PayPal, S-1 Filing with the S.E.C., September 28, 2001: http://www.nasdaq.com/markets/ipos/filing.ashx?filingid=1557068.

59. Jackson, *The PayPal Wars*, p. 189.

60. Ibid., p. 212.

61. Margaret Kane, "EBay Picks up PayPal for $1.5 Billion," Cnet, August 18, 2002.

62. Jackson, *The PayPal Wars*, p. 228.

63. Brian Caulfield and Nicole Perlroth, "Life After Facebook," *Forbes*, January 26, 2011.

64. Andy Greenberg and Ryan Mac, "How a 'Deviant' Philosopher Built Palantir, a CIA-Funded Data-Mining Juggernaut," *Forbes*, August 14, 2013.

65. Peter Thiel, "Ask Me Anything," Reddit, September 11, 2014: https://www.reddit.com/r/IAmA/comments/2g4g95/peter_thiel_technology_entrepreneur_and_investor/.

66. See Timothy B. Lee, "Peter Thiel Thought About the Election like a Venture Capitalist," *Vox*, November 11, 2016.

67. Casey Newton, "Mark Zuckerberg Defends Peter Thiel's Trump Ties in Internal Memo," *The Verge*, October 19, 2016.

68. George Packer, "No Death, No Taxes," *The New Yorker*, November 28, 2011.

69. Peter Thiel with Tyler Cowen, "Conversations with Tyler."

70. Nathan Ingraham, "Larry Page Wants to 'Set Aside a Part of the World' for Unregulated Experimentation," *The Verge*, May 15, 2013.

71. Peter Thiel, "The Education of a Libertarian," *Cato Unbound* blog, April 13, 2009.

72. Thiel, *Zero to One*, p. 24.

73. Ibid., pp. 24–25.

74. Farhad Manjoo, "Why Facebook Keeps Beating Every Rival: It's the Network, of Course," *The New York Times*, April 19, 2017.

75. Thiel, *Zero to One*, p. 32.

76. Max Levchin, *The Charlie Rose Show*.

77. Thiel, *Zero to One*, p. 26.

78. *Time* magazine cover, February 19, 1996.

79. Jeffrey M. O'Brien, "The PayPal Mafia," *Fortune*, November 26, 2007.

8. Reid Hoffman et al.

1. Gary Rivlin, "If You Can Make It in Silicon Valley, You Can Make It . . . in Silicon Valley Again," *The New York Times Magazine*, June 5, 2005, p. 64.

2. Internet Live Stats, "United States Internet Users, 2000–2016": http://www.internetlivestats.com/internet-users/us/.

3. Internet Live Stats, "Internet Users," http://www.internetlivestats.com/internet-users/.

4. D. Steven White, "U.S. E-Commerce Growth: 2000–2009," August 20, 2010: http://dstevenwhite.com/2010/08/20/u-s-e-commerce-growth-2000-2009/.

5. Scott Laningham, "DeveloperWorks Interviews: Tim Berners-Lee," IBM, August 22, 2006: http://www.ibm.com/developerworks/podcast/dwi/cm-int082206txt.html.

6. John Cloud, "The YouTube Gurus," *Time*, December 25, 2006.

7. Matt Richter, "Napster Appeals an Order to Remain Closed Down," *The New York Times*, July 13, 2001, p. 4.

8. Katie Hafner, "We're Google. So Sue Us," *The New York Times*, October 23, 2006, p. C1.

9. Jeffrey Rosen, "Inconspicuous Consumption," *The New York Times Book Review*, November 27, 2011, p. 18.

10. Miguel Helft, "It Pays to Have Pals in Silicon Valley," *The New York Times*, October 17, 2006, p. C1.

11. Rivlin, "If You Can Make It in Silicon Valley."

12. Eric M. Jackson, *The PayPal Wars: Battles with eBay, the Media, the Mafia, and the Rest of Planet Earth*, Washington, D.C.: WND Books, 2012, p. 24.

13. Ibid.

14. Reid Hoffman and Ben Casnocha, *The Start-up of You: Adapt to the Future, Invest in Yourself, and Transform Your Career*, New York: Crown Business, 2012, pp. 159–161.

15. Thomas Friedman, "The World Is Flat: A Brief History of the 21st Century," New York: Farrar, Straus and Giroux, 2005, p. 75.

16. Hoffman and Casnocha, *The Start-up of You*, p. 15.

17. Ibid., p. 19.

18. Ibid., p. 36.

19. Ibid., p. 239 (footnote 4, chapter 1).

20. Packer, "No Death, No Taxes."

21. Ibid., p. 19.

9. Jimmy Wales

1. Ted Greenwald, "How Jimmy Wales' Wikipedia Harnessed the Web as a Force for Good," *Wired*, March 19, 2013.

2. Bomis Sign-Up Page, accessed through the Wayback Machine, May 8, 1999: http://web.archive.org/web/19990508174513/http://my.bomis.com/member/signup.

3. Stacy Schiff, "Know It All," *The New Yorker*, July 31, 2006; Katherine Mangu-Ward, "Wikipedia and Beyond," *Reason*, June 2007.

4. See Zach Schwartz, "An Interview with the Founder of Wikipedia," Zach Two Times Blog, November 19, 2015: http://zachtwotimes.blogspot.com/2015/11/an-interview-with-founder-of-wikipedia.html.

5. Larry Sanger, "The Early History of Nupedia and Wikipedia: A Memoir," *Slashdot*, April 18, 2005: https://features.slashdot.org/story/05/04/18/164213/the-early-history-of-nupedia-and-wikipedia-a-memoir.

6. Terry Foote, e-mail with author, July 19, 2016.

7. Schwartz, "An Interview with the Founder of Wikipedia."

8. Wikipedia: Multilingual ranking January 2002: https://en.wikipedia.org/wiki/Wikipedia:Multilingual_ranking_January_2002.

9. Larry Sanger, "What Wikipedia Is and Why It Matters," talk to Stanford University Computer Systems Laboratory EE380 Colloquium, January 16, 2002: https://meta.wikimedia.org/wiki/Wikipedia_and_why_it_matters.

10. Larry Sanger, "The Early History of Nupedia and Wikipedia, Part II," *Slashdot*, April 19, 2005: https://slashdot.org/story/05/04/19/1746205/the-early-history-of-nupedia-and-wikipedia-part-ii.

11. Terry Foote, e-mail with author, July 22, 2016.

12. Larry Sanger, "Announcement About My Involvement in Wikipedia and Nupedia," February 13, 2002: https://meta.wikimedia.org/w/index.php?title=Announcement_about_my_involvement_in_Wikipedia_and_Nupedia--Larry_Sanger&action=history.

13. Nathaniel Tkacz, "The Spanish Fork: Wikipedia's Ad-Fuelled Mutiny," *Wired UK*, January 20, 2011.

14. Terry Foote e-mail, July 22, 2016.

15. "Wikia Continues Global Expansion with $15 Million in D-Round Funding," *PR Newswire*, August 27, 2014: http://www.prnewswire.com/news -releases/wikia-continues-global-expansion-with-15-million-in-d-round -funding-272899031.html.

16. Wikimedia Foundation Inc., "Financial Statements," June 30, 2016 and 2015 (With Independent Auditors Report Thereon): https://upload.wikimedia .org/wikipedia/foundation/4/43/Wikimedia_Foundation_Audit_Report_ -_FY15-16.pdf.

17. "Wikipedia Founder Jimmy Wales Responds," *Slashdot*, July 28, 2004: https://slashdot.org/story/04/07/28/1351230/wikipedia-founder-jimmy -wales-responds.

18. Jimmy Wales, "Keep Wikipedia Free": https://wikimediafoundation .org/wiki/Keep_Wikipedia_Free.

19. Sarah Mitroff, "Craig Newmark Sits at the Top and Bottom of Craigslist," *Wired*, July 16, 2012: https://www.wired.com/2012/07/craig-newmark/.

10. Mark Zuckerberg

1. Michael M. Grynbaum, "Mark E. Zuckerberg '06: The Whiz Behind Thefacebook.com," *The Harvard Crimson*, June 10, 2004.

2. Joseph Weizenbaum, "Computer Power and Human Reason: From Judgment to Calculation," New York: W.H. Freeman and Company, 1976, p. 115.

3. Jose Antonio Vargas, "The Face of Facebook," *The New Yorker*, September 20, 2010.

4. Designing Media, "Designing Media: Mark Zuckerberg Interview" (2010), Zuckerberg Transcripts, Paper 73: http://dc.uwm.edu /zuckerberg_files_transcripts/73.

5. Synapse promotional copy found on Myce.com, April 24, 2003: http: //www.myce.com/news/Intelligent-MP3-player-plays-the-right-song-at-the -right-moment-5776/.

6. "Machine Learning and MP3s," *Slashdot*, April 21, 2003: https://news .slashdot.org/story/03/04/21/110236/machine-learning-and-mp3s.

7. S. F. Brickman, "Not-so-Artificial Intelligence," *The Harvard Crimson*, October 23, 2003.

8. Mark Zuckerberg interview, Y Combinator, Startup School 2013, October 25, Zuckerberg Transcripts, Paper 160: http://dc.uwm.edu /zuckerberg_files_transcripts/160.

9. Mark Zuckerberg interview, Idea to Product Latin America, October 13, 2009, Zuckerberg Transcripts, Paper 92: http://dc.uwm.edu /zuckerberg_files_transcripts/92.

10. Mark Zuckerberg interview, Y Combinator, Startup School 2012, October 20, Zuckerberg Transcripts, Paper 161: http://dc.uwm.edu /zuckerberg_files_transcripts/161.

11. Zuckerberg, Y Combinator (2013).

12. Bonnie Goldstein, "The Diaries of Facebook Founder," *Slate*, November 30, 2007.

13. Bari M. Schwartz, "Hot or Not? Website Briefly Judges Looks," *The Harvard Crimson* November 4, 2003.

14. Luke O'Brien, "Poking Facebook," *02138*, November/December 2007, accessed via Wayback Machine: https://web.archive.org/web /20080514021019/http://www.02138mag.com/magazine/issue/58.html.

15. Zuckerberg, Y Combinator (2013).

16. The Harvard Crimson Staff, "M*A*S*H," *The Harvard Crimson*, November 6, 2003.

17. Matthew Shaer, "The Zuckerbergs of Dobbs Ferry, New York," *New York Magazine*, May 6, 2012.

18. Mark Zuckerberg, interview for Justin.tv's Startup School 2010, October 19, Zuckerberg Transcripts, Paper 35: http://dc.uwm.edu /zuckerberg_files_transcripts/35

19. See Richard Pérez-Peña, "To Young Minds of Today, Harvard Is the Stanford of the East," *The New York Times*, May 30, 2014, p. A1.

20. Ibid.

21. See O'Brien, "Poking Facebook," for the history of the development of Harvard Connection and thefacebook.com.

22. Grynbaum, "Mark E. Zuckerberg '06."

23. Alan J. Tabak, "Hundreds Register for New Facebook Website," *The Harvard Crimson*, February 9, 2004.

24. O'Brien, "Poking Facebook."

25. Brad Stone, "ConnectU's 'Secret' $65 Million Settlement with Facebook," *The New York Times*, Bits blog, February 10, 2009.

26. Matt Welsh, "In Defense of Mark Zuckerberg," Volatile and Decentralized blog, October 10, 2010: http://matt-welsh.blogspot.com/2010/10/in -defense-of-mark-zuckerberg.html.

27. John Cassidy, "Me Media," *The New Yorker*, May 15, 2006.

28. Tabak, "Hundreds Register for New Facebook Website."

29. Bianca Bosker, "Facebook's Mark Zuckerberg, Barefoot with Beer: 2005 Interview Reveals CEO's Doubts (VIDEO)," *The Huffington Post*, August 11, 2011, http://www.huffingtonpost.com/2011/08/11/facebook -mark-zuckerberg-2005-interview_n_924628.html; Transcript via Zuckerberg Transcripts, Paper 56: http://dc.uwm.edu/zuckerberg_files_transcripts /35.

30. Henry Blodget, "Mark Zuckerberg on How Facebook Became a Business" (2010), Zuckerberg Transcripts, Paper 8: http://dc.uwm.edu /zuckerberg_files_transcripts/8.

31. Welsh, "In Defense of Mark Zuckerberg."

32. Grynbaum, "Mark E. Zuckerberg '06."

33. Cassidy, "Me Media."

34. Zuckerberg, Y Combinator (2012): "Q. Why did you choose ones that had school specific social network?

Mark Zuckerberg: Because I wanted to—

Q: Because they could become competitors?

Mark Zuckerberg: Well I wanted to go to the schools that I thought would be the hardest for us to succeed at."

35. Zuckerberg, Y Combinator (2013).

36. Blodget, "Mark Zuckerberg on How Facebook Became a Business."

37. Brian Caulfield and Nicole Perlroth, "Life After Facebook," *Forbes*, January 26, 2011.

38. Zuckerberg, Y Combinator (2013).

39. David Kushner, "The Baby Billionaires of Silicon Valley," *Rolling Stone*, November 16, 2006.

40. Stanford University, "James Breyer/Mark Zuckerberg Interview, October 26, 2005, Stanford University" (2005), Zuckerberg Transcripts, Paper 116: http://dc.uwm.edu/zuckerberg_files_transcripts/116.

41. Zuckerberg, Y Combinator (2013).

42. Dealbook, "Tracking Facebook's Valuation," *The New York Times*, February 1, 2012.

43. Stanford University, "James Breyer/Mark Zuckerberg Interview."

44. David Kushner, "Being Mark Zuckerberg," IEE E Spectrum blog, September 16, 2010, http://spectrum.ieee.org/tech-talk/geek-life/profiles /being-mark-zuckerberg.

45. Biana Bosker, "Facebook's Mark Zuckerberg, Barefoot With Beer: 2005 Interview Reveals CEO's Doubts (VIDEO)."

46. Mark Zuckerberg interview, Y Combinator, Startup School 2011, October 30. Zuckerberg Transcripts, Paper 76: http://dc.uwm.edu /zuckerberg_files_transcripts/76.

47. Stanford University, "James Breyer / Mark Zuckerberg Interview."

48. Zuckerberg, Y Combinator (2013).

49. Tad Friend, "Tomorrow's Advance Man," *The New Yorker*, May 18, 2015.

50. Peter Thiel with Blake Masters, *Zero to One: Notes on Startups, or How to Build the Future*, New York: Penguin Random House, 2014, p. 80.

51. Zuckerberg, Y Combinator (2011).

52. Sarah Lacy, *Once You're Lucky, Twice You're Good*, New York: Gotham Books, 2008, p. 182.

53. Zuckerberg, Y Combinator (2011).

54. Mark Zuckerberg, interview for Justin.tv's Startup School 2010.

55. Evelyn M. Rusli, "Profitable Learning Curve for Facebook CEO Mark Zuckerberg," *The Wall Street Journal*, January 5, 2014.

56. Gideon Lewis-Kraus, "The Great AI Awakening," *The New York Times Magazine*, December 14, 2016.

57. Mark Zuckerberg and Priscilla Chan, "A Letter to Our Daughter," December 1, 2015: https://www.facebook.com/notes/mark-zuckerberg/a-letter-to-our-daughter/10153375081581634/.

58. Mark Zuckerberg, "Our Commitment to the Facebook Community," note to Facebook's Facebook page, November 29, 2011: https://www.facebook.com/notes/facebook/our-commitment-to-the-facebook-community/10150378701937131/.

59. Video interview with Steve Jobs, "Memory and Imagination: New Pathways to the Library of Congress," directed by Julian Krainin and Michael R. Lawrence: https://www.youtube.com/watch?v=ob_GX50Za6c.

60. Zuckerberg, Y Combinator (2012).

61. Robin Dunbar, "You've Got to Have (150) Friends," *The New York Times*, December 25, 2010, p. WK15.

62. Ibid.

63. Ibid.

64. See Dale Russakoff, *The Prize: Who's in Charge of America's Schools?*, New York: Houghton Mifflin Harcourt, 2015.

65. Zuckerberg and Chan, "A Letter to Our Daughter."

66. Ibid.

67. Mark Zuckerberg, video, May 4, 2015, available at Mark Zuckerberg's Facebook page.

68. See Mahesh Murthy, "Internet.org Is Just a Facebook Proxy Targeting India's Poor," *Firstpost.com*, April 17, 2015.

69. Savetheinternet.in Coalition, "Dear Mark Zuckerberg, Facebook Is Not, and Should Not Be the Internet," *Hindustan Times*, April 17, 2015.

70. Adi Narayan, "Andreessen Regrets India Tweets; Zuckerberg Laments Comments," Bloomberg.com, February 10, 2016.

71. Mark Zuckerberg, video, May 4, 2015, available at Mark Zuckerberg's Facebook page.

72. Zuckerberg and Chan, "A Letter to Our Daughter."

The Future

1. George Packer, "No Death, No Taxes," *The New Yorker*, November 28, 2011.

2. John McCarthy, "Artificial Intelligence and Creativity," Century 21 lecture, January 30, 1968, audio file in KZSU Collection, Stanford Archive of Recorded Sound, Stanford University Libraries.

3. Peter Thiel, "Ask Me Anything," Reddit, September 11, 2014: https://www.reddit.com/r/IAmA/comments/2g4g95/peter_thiel_technology _entrepreneur_and_investor/.

4. Douglas Hofstadter, email re: 2009 SAIL Reunion, October 29, 2009, accessed via Regrets Web page: https://web.stanford.edu/~learnest/spin /regrets.htm.

5. Peter Thiel on the Future of Innovation with Tyler Cowen, "Conversations with Tyler," Mercatus Center, April 6, 2015: https://medium .com/conversations-with-tyler/peter-thiel-on-the-future-of-innovation -77628a43c0dd#.bav03wzih.

6. Mark Zuckerberg and Priscilla Chan, "A Letter to Our Daughter," December 1, 2015: https://www.facebook.com/notes/mark-zuckerberg/a -letter-to-our-daughter/10153375081581634/.

7. Mark Zuckerberg, "Building Global Community," Facebook, February 16, 2017, https://www.facebook.com/notes/mark-zuckerberg/building -global-community/10103508221158471/?pnref=story.

8. Daniel T. Rodgers, *The Age of Fracture*, Cambridge, MA: Harvard University Press, 2011, p. 194, discussing Michael Walzer, *Spheres of Justice: A Defense of Pluralism and Equality*, New York: Basic Books, 1983.

A Note to the Reader

1. Diane Brady, "In Ben Horowitz's New Book, Women Are Markedly Absent," *Bloomberg Businessweek*, March 12, 2014.

2. Marc Andreessen, post to Twitter, March 12, 2014.

INDEX

ABOUT THE AUTHOR

Noam Cohen covered the influence of the Internet on the larger culture for the *New York Times*, where he wrote the Link by Link column, beginning in 2007. He lives in Brooklyn with his family. This is his first book.